Lecture Notes in Control and Information Sciences

Edited by M. Thoma

For information about Vols. 1–42 please contact your bookseller or Springer-Verlag.

Lecture Notes in Control and Information Sciences

Edited by M. Thoma and A. Wyner

98

A. Aloneftis

Stochastic Adaptive Control Results and Simulations

Springer-Verlag Berlin Heidelberg GmbH

Series Editors
M. Thoma · A. Wyner

Advisory Board
L. D. Davisson · A. G. J. MacFarlane · H. Kwakernaak
J. L. Massey · Ya Z. Tsypkin · A. J. Viterbi

Author
Alexis Aloneftis
Korytsas 2
St. Andreas, Nicosia 171
Cyprus

ISBN 978-3-540-18055-5   ISBN 978-3-540-47726-6 (eBook)
DOI 10.1007/978-3-540-47726-6

Library of Congress Cataloging in Publication Data
Aloneftis, A. (Alexis)
Stochastic adaptive control.
(Lecture notes in control and information scienes; 98)
Abstract in French.
Bibliography: p.
1. Adaptive control systems. 2. Stochastic systems.
I. Title. II. Series.
TJ217.A45   1987   629.8'36   87-16411

© Springer-verlag Berlin Heidelberg 1987
Originally published by Springer-Verlag Berlin Heidelberg New York in 1987.

Abstract

The theory of the direct method of self-tuning control of SISO time invariant (TI) ARMAX models is developed. Results are presented on parameter estimation and self-tuning control using the stochastic gradient, least squares, quasi-least squares, and modified least squares (MLS) algorithms. Computer simulations confirm stabilization and tracking, the MLS algorithm resulting in the best performance.

The control of SISO time varying (TV) ARMAX models is examined and general controller design considerations discussed. Adaptive minimum variance and maximum likelihood controllers for TV ARX systems whose parameters are a finite state homogeneous Markov chain are developed. Finally, a new approach for the analysis of TV ARMAX systems is presented using a related state space model, the state being a Markov process with stationary transition probabilities. Simulations of algorithms for TV systems follow.

Résumé

La théorie de la méthode directe de la commande auto-ajustable pour des modèles ARMA(X) unidimensionelles stationnaires est developpé. Des résultats sont présentés sur l'estimation des paramètres et la commande auto-ajustable. Ils font appel à les algorithmes du gradient stochastique, des moindres carrés, des quasi-moindres carrés, et des moindres carrés modifiés (MLS). Les simulations par ordinateur confirment la stabilisation et la localisation des algorithmes, l'algorithme des MLS donnant la meilleur performance.

Par ailleurs, la commande de modèles ARMA(X) unidimensionels non-stationnaires est étudiée et des considérations générales pour la conception de contrôleurs sont examinées. Les commandes adaptatives à variance minimale et à vraisemblance maximale sont developpées pour des systèmes ARX non-stationnaires dont les paramètres sont une chaîne de Markov (faible) homogène à nombre fini d'états. Enfin, une nouvelle méthode pour l'analyse des modèles ARMAX non-stationnaires est présentée. Elle se sert d'un modèle d'état connexe pour lequel l'état est un processus de Markov à probabilités de transition stationnaires. Des simulations d'algorithmes pour de tels systèmes s'ensuivent.

Preface

This work was written with two objectives in mind. First, to present in a motivated way the direct method of parameter self-tuning control as applied to linear single-input-single-output time invariant systems subject to random disturbances, and its extension to the control of time varying systems. Second, to present computer simulations of most of the algorithms discussed and compare theoretical with experimental results. It is felt that most of the ideas which arise in the study of adaptive control are intuitively appealing and philosophical in content. It is therefore hoped that the presentation of various ideas and concepts without commitment to any one set in particular will provide for thought provoking and interesting reading.

In an effort to enhance the presentation of the results, the following conventions and notational devices have been used and should be noted prior to reading this study.

• Appendix A contains a set of technical assumptions used throughout this study. Reference to it is made via a capital letter and usually followed by a number, both enclosed in parentheses. The capital letter is not necessarily A. For example, (W5) and (T) are both references to Appendix A.

• References to Appendix B and C start with the capital letters B and C respectively and are followed by a number. For example, (B1.1) and (B3) are both references to Appendix B while (C1) is a reference to Appendix C.

• Certain equation numbers end in a small alphabetic character (e.g. 2.1a), which makes them part of a group of equations referred to by its equation number (e.g. 2.1).

• The symbol ■ is used to define the end of a titled group of statements, be it a theorem, a corollary, a definition, or an example.

• Vectors and matrices are respectively denoted by lowercase and uppercase boldface characters.

• As most quantities are random variables, the generic argument ω has been omitted except in certain cases where such a dependence needs to be emphasized.

• The word process shall be taken to mean stochastic process.

All quantities, abbreviations, and further conventions are defined in a separate section prior to Chapter 1.

Acknowledgements

I am indebted to Prof. Peter E. Caines for his guidance and personal involvement during the course of the research. His enthusiasm, friendship, and encouragement during periods of despair have greatly contributed to my understanding of the research process and to maintaining my sanity. I am grateful to him for his financial support.

Special thanks for their friendly support and for discussions on technical matters go to Youssef Ghoneim, Yong Zhou, Yuan Zhong, Mike Sabourin, Elabed Saud, Michel Habib, Mark Readman, Olivier Gomart, Stelios Louisides, Lawrence Joseph, Carey Levinton, Shari Grabelle, Giuliana Minardi, and Maria Tsakalis.

In particular, I am happy to acknowledge my gratitude to my close friend Patrick Aboussouan (ybbuh) for his moral support, companionship, and love of life; to Christian Hugues Houdré for his friendship and his mathematical wisdom which he so generously shared with me; to Sean Meyn for his collaboration, constant encouragement, and warm and friendly humour; to Hans Wolter Wehn for his support and the many philosophical discussions he shared with me; to the Vlahos for their love, concern, and Sunday talks; to Demetri and Kalliopi Pavlides and to my brother Serge for their love and support. Lastly and most importantly I wish to extend my gratitude to my parents for the greatest gift I have received so far: my education.

The computer simulations found herein would not have been possible without the generous support of the McGill University Computing Centre. The facilities and consultation they have provided are gratefully acknowledged. Thanks also go to Jacek Slaboszewicz for welcoming and aiding me when in "computer despair".

The manuscript was expertly and expediently typeset by Mrs. Mindle Levitt. Her humorous and caring attitude is hereby acknowledged. Many thanks to Pierre Parent for his expert help in typesetting the manuscript.

Table of Contents

List of Figures

Notation, Conventions, and Abbreviations

Notation

t	an integer representing the discrete time parameter.
$\widehat{e}(\cdot)$	an estimate of a quantity $e(\cdot)$.
$\boldsymbol{\vartheta}, \boldsymbol{\vartheta}^\circ, \boldsymbol{\vartheta}(\cdot)$	generic symbol of a parameter vector. Defined in context.
$\boldsymbol{\varphi}, \boldsymbol{\varphi}(\cdot)$	generic symbol of a regression vector. Defined in context.
$\lambda_{\min}(\mathbf{A}), \lambda_{\max}(\mathbf{A})$	the minimum and maximum eigenvalues of the matrix $\mathbf{A}$ respectively.
$\mathcal{F}$	the generic symbol for a σ-algebra.

Conventions

$z^k x(t)$	a right shift of k integer units for any discrete time variable $x(\cdot)$.
$\boldsymbol{\vartheta}^T$	the transpose of the vector $\boldsymbol{\vartheta}$.
$\dim \boldsymbol{\vartheta}$	the dimension of the vector $\boldsymbol{\vartheta}$.
x_1^t	the set $\{x(1), \cdots, x(t)\}$.
$x \in \mathcal{F}$	x is measurable with respect to $\mathcal{F}$.
$\sigma\{\cdots\}$	the σ-algebra generated by the quantities within braces.
$\Omega \backslash \Gamma$	the complement of a set Γ in Ω.
$x \perp\!\!\!\perp y$	x and y are statistically independent.
$P(\cdot)$	the probability of the set in brackets.
$\mathbb{1}_S(x)$	the characteristic function of the set S.
$E[y]$	the expectation of a random variable y.
$E[y\|\mathcal{F}]$	the conditional expectation of y with respect to the σ-algebra $\mathcal{F}$.
$\sim$	is distributed as.
$\mu \prec \nu$	μ is absolutely continuous with respect to ν.
$\Re(\Re^n)$	the set of real numbers (the set $\Re \times \Re \times \cdots \times \Re$).
$\Re^+$	the set of non-negative real numbers.
$\mathbf{C}$	the set of complex numbers.
$Z(Z^+)$	the set of integers (the set of non-negative integers).
$\aleph$	the set of natural numbers.
$\triangleq \ (\underset{\nabla}{=})$	is defined equal to (is denoted by).

Abbreviations

ARMA(X)	AutoRegressive Moving Average (with eXogenous inputs).
ASTE	Averaged Squared Tracking Error.
a.a.(ν) x	for almost all x with respect to the measure ν.
a.s.	almost surely.
CDC	Continually Disturbed Controls.
i.i.d.	independent and identically distributed.
LQ(G)	Linear Quadratic (Gaussian).
LTI	Linear Time Invariant.
LTV	Linear Time Varying.
ML	Maximum Likelihood.
MLS	Modified Least Squares.
MV	Minimum Variance.
$N(\mu, \sigma^2)$	the normal distribution of mean μ and variance σ^2.
QLS	Quasi-Least Squares.
RLS	Recursive Least Squares.
r.v.	random variable.
SG	Stochastic Gradient.
SISO	Single Input Single Output.
SPR	Strictly Positive Real.
s.t.	such that.
$V(a, b)$	the uniform distribution in the interval (a, b).
w.p.	with probability.
w.r.t.	with respect to.

T he process of adaptation is akin to that of human learning, no doubt a complex process. It is not surprising then that, in the context of Automatic Control, it is difficult to find less complex and more vague concepts than that of adaptation. Definitions of adaptive control ([7, Bar-Shalom and Gershwin], [68, Tsypkin]) range from the appealing picture of *a hierarchical structure of feedback* to the pragmatic one of *a method of approach when the exact formulation as a non-linear control problem is deemed too complex.* Also enticing is the view of *an effort to simulate the human learning system*, a view which links adaptive control to artificial intelligence. Whichever point of view one decides to adopt, it is instructive to compare adaptive control with "ordinary" feedback control. One of the main purposes of the latter is to safeguard the performance of the overall system against plant parameter variations and variations in the environment caused by external disturbances. The general idea of increasing the feedback gain will not always succeed in stabilizing the system to be controlled and can in fact destabilize it. Without an increase in a priori information it is impossible to increase the feedback gain and not affect the stability of the system. This immediately poses the question of what distinguishes adaptive control from plain feedback control, but also provides the clue that adaptation is intimately connected with the lack of a priori information. A view which encompasses many of the aforementioned ideas above is that *adaptation is the estimation of information that, if given a priori, would enable feedback regulation* [15, Caines]. As was remarked in [19, Caines and Chen], "if one wishes to relax presently required a priori information, one is forced to use some further level of adaptation". This view results in some kind of hierarchy of feedback and points to the iterative nature of adaptive control. The distinction between feedback and adaptation is not clear; however, it can be said that, in a certain sense, feedback copes with ignorance while adaptation removes it.

It must be noted from the outset that for the problems addressed herein, it is the lack of knowledge of the system model parameters, as opposed to some other quantity, which makes the problem an adaptive one. Consequently, from now on, the adaptive control problem shall be understood to mean the parameter adaptive control problem.

Adaptive control problems are classified as Bayesian and Non-Bayesian problems. Their essential difference is based on how the unknown parameter ϑ enters the problem formulation. In the *Bayesian framework*, ϑ is a random quantity with some given initial (prior) distribution, the goal being to minimize a cost criterion. Typically, this criterion is the expected value of a function which depends on the input and state of the system, the crucial point being that this expectation is taken with respect to the random system behaviour *as well as* the random behaviour of ϑ via its prior distribution. This approach leads to non-linear estimation and control problems and will not be discussed. In the *Non-Bayesian framework*, the one adopted here, ϑ is simply a non-random quantity lying in some specified set.

The ideas of dual effect, neutrality, separability, and certainty equivalence are found throughout the literature, and although definitions of them have been given, they are used somewhat loosely. *Duality* refers to the ability of the controller to affect the uncertainty in the knowledge of the state of a system. The term dual is used to refer to the dual role of the controller: to control a system and to influence learning [37, Feldbaum]. These roles of the controller are in conflict since good control requires knowledge of the plant while determining the plant parameters requires probing control actions. As pointed out by Shannon, this duality is closely related to that of past and future. While it is possible to know the past, one cannot control it. Similarly, it is possible to control the future but one cannot know it. Problems where the controller cannot affect the uncertainty of the state of a system are called *neutral*. An example is the LQG optimal control problem under certain often used assumptions. *Separability* is said to hold if the functional dependence of the optimal control — obtained for the corresponding deterministic system with complete state observation — on the state can be replaced by *some* functional dependence on the best estimate of the state while retaining optimality. Further, *certainty equivalence* is said to hold if the functional dependence of the deterministic optimal control (mentioned above) on the state is unchanged when state estimates replace the true state, and optimality is still retained. Clearly, certainty equivalence implies separability but not vice versa. It is well known that the LQG optimal control problem has the certainty equivalence property (and hence separability) and is often quoted to demonstrate or loosely define separability (see e.g. [43, Goodwin and Sin]). In [61, Patchell and Jacobs], it is conjectured that neutrality (or the lack of a "learning" capability) is a sufficient condition for certainty equivalence and indeed, [8, Bar-Shalom and Tse] prove that, for the LQG problem, (partial) neutrality is actually equivalent to certainty equivalence. This is not surprising since certainty equivalence controllers are optimal so that no further information, such as state uncertainty, can improve their performance.

When a given control problem is known to have the dual effect, one is still faced with the decision of whether to use the "learning" ability of the controller to hopefully improve its performance, given that this results in very complex control problems. When it is not known whether or not a problem has the dual effect, one usually pretends that the problem *is* certainty equivalent and proceeds. Remarkably, this ad hoc strategy, now known as the *self-tuning strategy*, is found to yield asymptotically optimal results and the controller is often quoted as a certainty equivalence controller or one designed on the certainty equivalence principle. It is the only type of controller considered in this thesis in the setting of ARMAX system models with time invariant and time varying parameters.

A dual controller (one which uses its "learning" capability), must, by definition, use information about future as well as present and past observations such as predicted uncertainty of future parameters — this of course does not violate the requirement that the controls be non-anticipative. Such controllers are called *actively adaptive* (see [8, Bar-Shalom and Tse]), while those which use information about present and past observations only, are named *passively adaptive*. Passively adaptive controllers which, in addition to using the recently obtained observations, also use their estimated uncertainty to influence the control decision, are called *cautious*. The ideas of caution, active or passive adaptation, and others mentioned above, have resulted in very few concrete results. Nevertheless, the descriptive flavour which they bring to the interpretation of the theme of adaptation is felt to be valuable and helpful.

There is, as in life, a compromise to be made between control, caution, and learning. In all cases above, a name has been given to the controller based on its ability to perform these tasks and it is felt that this point of view is the most appropriate one. A second point of view, and one which, for simplicity, is adopted in this thesis, is that of naming the controller adaptive or self-tuning depending, respectively, on whether the plant to be controlled is time varying or not. This terminology is likely to coincide with one's first thought of adaptive control.

This study presents the self-tuning strategy for discrete time SISO ARMAX systems the goal being to track a given arbitrary deterministic reference sequence in a mean square sense. Model following, the tracking of a sequence generated by a specified model, and regulation, the tracking of a constant sequence, are special cases of the tracking problem. The above goal is also expanded to include simultaneous consistent parameter estimation. The so-called direct method of adaptive control, where the regulator parameters are directly estimated, is used throughout. The major results found scattered in various journals and reports are presented, and computer simulations of most of the control algorithms covered are provided. The objective is to concentrate simulations of adaptive control into one report in order to display their comparative behaviour, and to indicate the pitfalls, usefulness in applications, and new directions for investigation.

The next chapter contains an exposition of parameter estimation and self-tuning control of LTI systems, followed by a set of well known, as well as recent, results. Chapter 3 presents computer simulations of most of the algorithms of Chapter 2. Chapter 4 shifts to a time varying setting and presents the few available results together with a new approach to the stochastic adaptive control problem. Chapter 5 provides insightful simulations of some results of Chapter 4, while Chapter 6 concludes the study.

Self-Tuning Control of Systems
with Random Disturbances

2.1 Introduction

This chapter examines the control of constant parameter systems whose parameters are assumed to be unknown. This lack of knowledge is what motivates the common use of the label "adaptive" or "parameter adaptive" although it is felt that the adjectives "self-tuning" or "parameter self-tuning" are more suggestive since they portray the situation of a controller which tries to tune to an unknown *time invariant* system. The dictionary definition of adaptation *to alter so as to fit for a new use* encompasses the situation described above but does not allow one to distinguish between controllers that have the ability to change as the environment changes and those that do not. Consequently, the term adaptive will be reserved in this thesis to describe controllers designed for *time varying* systems.

The structure of such self-tuning controllers can be easily visualized when their performance is viewed as the combination of two steps: learning and acting. The learning step corresponds to the estimation of a set of parameters of an a priori chosen model based on previous observations. The action step uses the latest available estimate of the parameters to compute and apply an input (usually called the control) which will contribute towards the achievement of a pre-specified objective. The choice of parameter estimators and controllers is dictated by that of criteria for each purpose. The problem at hand is fundamentally a control and not an estimation problem, and as such, the control criterion (MV, LQ, etc.) is fixed. On the other hand, there is a vast choice of parameter estimation algorithms that can be used. Certain of them originate from the optimization of various criteria, while others are well motivated variations of particular schemes. Here, the presentation is confined to those estimation algorithms which have been extensively studied in the context of adaptive control. It is tempting to think that the parameter estimation algorithms which have found application in adaptive control, did so because of some of their desirable convergence properties. It is important to note however, that properties such as consistency can easily be lost when these algorithms are placed in a feedback configuration, see e.g. [5, Åström and Wittenmark]. In fact, most convergence proofs require some form of stability which cannot be assumed a priori in the adaptive problem.

The object of this chapter is to present the most recent relevant results on self-tuning control and parameter estimation for the tracking of an arbitrary reference sequence. All subsequent discussions in this chapter are based on an assumed model for the system to be controlled. This is taken to be the well known ARMAX model (2.1) below, which will also be referred to in the sequel as the system model. Such models are equivalent to stochastic LTI state space models [15, Caines].

The System Model

$\{u(t)\}, \{y(t)\}$, and $\{w(t)\}$, are scalar, real valued, input, output, and disturbance processes respectively which satisfy

$$A(z)y(t) = B(z)u(t-d) + C(z)w(t), \quad t \geq 1, \quad d \geq 1, \qquad (2.1a)$$

where

$$A(z) = 1 + a_1 z + \cdots + a_n z^n, \qquad (2.1b)$$

$$B(z) = b_0 + b_1 z + \cdots + b_m z^m, \quad b_0 \neq 0, \qquad (2.1c)$$

and

$$C(z) = 1 + c_1 z + \cdots + c_\ell z^\ell. \qquad (2.1d)$$

The initial condition

$$x_0 = [y(0)\cdots y(1-n), u(-d)\cdots u(1-d-m), w(0)\cdots w(1-\ell)]^T \qquad (2.1e)$$

is assumed to be given. ∎

2.2 The Direct Method of Self-tuning Control

The class of minimum (error) variance (MV) controllers are those which, at each instant t, minimize

$$V(t) = E[(y(t) - y^*(t))^2] \tag{2.2}$$

where $\{y^*(t)\}$ is a given deterministic reference sequence. For the model (2.1), an input can only effect an output d time instants later, and hence minimizing $V(\cdot)$ at time t amounts to choosing an input $u(\cdot)$ at time $t - d$ which, d instants later, will produce an output $y(t)$ as close to $y^*(t)$ as possible. It is clear however that one cannot deduce the effect of the choice of $u(t - d)$ on $y(t)$ in an exact way — even when the system model parameters are known — and so one tries to predict this effect to the best of one's ability. Naturally, this involves use of the latest available information, namely, all available inputs and outputs. This information is accumulated in the non-decreasing sequence of observation σ-algebras

$$\mathcal{F}_0^y \subset \mathcal{F}_1^y \subset \cdots \subset \mathcal{F}_t^y$$

where

$$\mathcal{F}_0^y \triangleq \sigma\{y(1-n)\cdots y(0), u(1-d-m)\cdots u(1-d), w(1-\ell)\cdots w(0)\} \underset{\bigtriangledown}{=} \sigma\{x_0\}$$

and

$$\mathcal{F}_t^y \triangleq \sigma\{y(1-n)\cdots y(t), u(1-d-m)\cdots u(1-d), w(1-\ell)\cdots w(0)\}, \quad t \geq 1,$$
$$\underset{\bigtriangledown}{=} \sigma\{x_0, y(1)\cdots y(t)\}.$$

It is assumed throughout this chapter that $u(t) \in \mathcal{F}_t^y$ whence, via (2.1), one observes that

$$\mathcal{F}_t^y = \mathcal{F}_t, \quad t \geq 1,$$

where

$$\mathcal{F}_t \triangleq \sigma\{y(1-n)\cdots y(0), u(1-d-m)\cdots u(1-d), w(1-\ell)\cdots w(t)\}$$
$$\underset{\bigtriangledown}{=} \sigma\{x_0, w(1)\cdots w(t)\}.$$

8

It is well known that in the class of $\mathcal{F}_{t-d}^y$ measurable predictors, the optimal (least squares) predictor of $y(t)$ given $\mathcal{F}_{t-d}^y$ is given by the conditional expectation

$$y^\circ(t|t-d) \triangleq E[y(t)|\mathcal{F}_{t-d}^y].\tag{2.3}$$

Under assumption (W1), a standard calculation (see e.g. [15, Caines]) gives

$$C(z)y^\circ(t|t-d) = G(z)y(t-d) + B(z)F(z)u(t-d), \quad t \geq 1,\tag{2.4}$$

with initial condition computable from x_0 and where

$$G(z) = g_0 + g_1 z + \cdots + g_{\bar{n}-1}z^{\bar{n}-1}, \quad \bar{n} = \max(\ell - d + 1, n),\tag{2.5a}$$

and

$$F(z) = 1 + f_1 z + \cdots + f_{d-1}z^{d-1}\tag{2.5b}$$

are the unique polynomials satisfying

$$C(z) = F(z)A(z) + z^d G(z).\tag{2.5c}$$

Note that such a factorization is possible (see e.g. [14, Burton]) and that the crucial fact that yields time invariant operators in (2.4) is that $x_0 \in \mathcal{F}_t^y$ for $t \geq 0$. Further, one also finds that the prediction error

$$v(t) \triangleq y(t) - y^\circ(t|t-d), \quad t \geq 1,\tag{2.6a}$$

satisfies

$$v(t) = F(z)w(t)\tag{2.6b}$$

and that by (W1), $(v(t), \mathcal{F}_t)$ is a martingale difference process. The predictor $y^\circ(t|t-d)$ given by (2.4) is optimal in the sense that

$$E[(y(t) - y^\circ(t|t-d))^2] \leq E[(y(t) - \tilde{y}(t|t-d))^2]$$

for *any* $\mathcal{F}_{t-d}^y$ measurable predictor $\tilde{y}(t|t-d)$ of $y(t)$. If in addition to (W1) assumption (W2) is imposed, the minimum mean square prediction error is given by

$$E[v^2(t)] = \left(1 + \sum_{i=1}^{d-1} f_i^2\right)\sigma_w^2 \triangleq \gamma^2.\tag{2.7}$$

It is now clear that setting $y^\circ(t|t-d)$ to $y^*(t)$ and solving for $u(t-d)$ via (2.4) will minimize (2.2). Such a solution exists since $b_0 \neq 0$, and the dependence of y° on u will be emphasized by writing y_u°. Then, to mathematically express what is said above, consider any deterministic sequence $\{y^*(t)\}$ specified for $t \geq 1$ and *arbitrary* for $t < 1$. Subtract $C(z)y^*(t)$ from both sides of (2.4) to get

$$C(z)(y_u^\circ(t|t-d) - y^*(t)) = G(z)y(t-d) + B(z)F(z)u(t-d) - C(z)y^*(t), \quad t \geq 1.$$

This equation may be rewritten in the following more compact form.

The Predictor Model

The optimal predictor $y_u^\circ(t|t-d)$ satisfies

$$C(z)(y_u^\circ(t|t-d) - y^*(t)) = \varphi^T(t-d)\vartheta^\circ - y^*(t), \quad t \geq 1, \tag{2.8a}$$

where

$$\varphi(t-d) = [y(t-d)\cdots y(t-d-\bar{n}+1), u(t-d)\cdots u(t-2d-m+1), -y^*(t-1)\cdots -y^*(t-\ell)]^T, \tag{2.8b}$$

and

$$\vartheta^\circ = [g_0 \cdots g_{\bar{n}-1}, (bf)_0 \cdots (bf)_{m+d-1}, c_1 \cdots c_\ell]^T, \tag{2.8c}$$

$(bf)_i$ being the coefficient of z^i in $B(z)F(z)$. ∎

Then if $u(t-d)$ is determined via the feedback scheme

$$\varphi^T(t-d)\vartheta^\circ = y^*(t), \quad t \geq 1, \tag{2.9}$$

the resulting control, denoted by $u_{mv}(t-d)$, will produce a closed loop system

$$C(z)(y_{u_{mv}}^\circ(t|t-d) - y^*(t)) = 0, \quad t \geq 1. \tag{2.10}$$

Notice that, at $t = 1$, the controls $u(1-d)\cdots u(\ell-d)$ to be solved from (2.9), as well as the initial conditions for $y_{u_{mv}}^\circ(t|t-d) - y^*(t)$ in (2.10) are arbitrary because they depend on the arbitrary quantities $y^*(1-\ell)\cdots y^*(0)$. The situation can be simplified if one chooses $y^*(1-\ell)\cdots y^*(0)$ so that

$$y^*(t) = y^\circ(t|t-d), \quad t = 1-\ell, \cdots, 0$$

thereby giving zero initial conditions for (2.10) and specifying the controls $u(1 - d), \cdots, u(\ell - d)$. In such a case, (2.10) implies that $y^*(t) = y^\circ(t|t - d)$ $\forall t \geq 1$, or using (2.6a), that

$$y(t) - y^*(t) = v(t), \qquad \forall t \geq 1. \tag{2.11}$$

Note that (2.6b) and (W1) imply that $\{v(t)\}$ is a zero mean process so that (2.2) actually expresses a variance and not just a second moment.

Then, from (2.7)

$$E[y(t) - y^*(t))^2] = \gamma^2, \qquad \forall t \geq 1, \tag{2.12}$$

and this is clearly the smallest achievable mean square tracking error with *any* causal (i.e. measurable w.r.t. $\mathcal{F}_t$) feedback $u(t)$. It is stressed that (2.11) is obtained because of the appropriate choice of $y^*(t)$ for $t < 1$. If such a choice is not possible (e.g. when the system model parameters are unknown, or, if one simply looks at the predictor model with arbitrary initial conditions as in [43, Goodwin and Sin]), then (2.11) should be replaced by

$$\lim_{t \to \infty} y(t) - y^*(t) - v(t) = 0$$

provided

$$C(z) = 0 \Rightarrow |z| > 1. \tag{2.13}$$

In this case, the effect of the initial conditions decays geometrically and the control given by (2.9) asymptotically approaches a minimum variance control. Condition (2.13) is implied by each of (M5), (M6), and (M7), one of which is always assumed in results of stochastic adaptive control. Such conditions may be relaxed at the expense of overparameterization (see [64, Shah and Franklin]).

As in (2.8), the parameters of the system model may be explicitly displayed by writing (2.1a) in vector form. Let

$$\varphi(t - 1) = [y(t - 1) \cdots y(t - n), u(t - 1) \cdots u(t - m - 1), w(t - 1) \cdots w(t - \ell)]^T \tag{2.14a}$$

and

$$\vartheta^\circ = [-a_1 \cdots - a_n, b_0 \cdots b_m, c_1 \cdots c_\ell]^T. \tag{2.14b}$$

Then (2.1a) can be rewritten as

$$y(t) = \varphi^T(t - 1)\vartheta^\circ + w(t). \tag{2.14c}$$

Remark: The argument t of $\varphi(t)$ denotes that the latter is measurable w.r.t. $\mathcal{F}_t$, i.e. it contains information up to and including the instant t. ■

The discussion so far has assumed the complete knowledge of the system model parameters and hence those of the predictor model. If these parameters are unknown, (2.8) and (2.14) suggest two similar control strategies. The first one is to estimate ϑ° of (2.14b), compute the corresponding estimate $\widehat{\vartheta}$ of ϑ° in (2.8) using the factorization (2.5), and then solve for the control $u(t-d)$ by substituting $\widehat{\vartheta}$ for ϑ° in (2.9) i.e. via

$$\varphi^T(t-d)\widehat{\vartheta}(t-d) = y^*(t). \tag{2.15}$$

This indirect or explicit method, involves the estimation of the $n+m+\ell+1$ system model parameters and the computation of the predictor model parameters for each iteration of a recursive implementation of this method. The alternative strategy is to estimate ϑ° in (2.8) directly and then solve for $u(t-d)$ in a similar way. This is known as the direct or implicit, method and involves the estimation of the $\bar{n} + m + \ell + d$ parameters of the predictor model at each iteration. While both methods implement the so-called self-tuning methodology they differ in some respects. The *indirect (explicit)* method *explicitly* estimates the system model parameters which *indirectly* determine the control law through equation (2.5). On the other hand, the *direct (implicit)* method *directly* determines the control law, which through (2.5), *implicitly* determines the system model parameters. There are, in general, more parameters to estimate in the direct method, but no extra computation is necessary since the primary objective is that of minimizing (2.2) and not that of estimating the system model parameters. Throughout this study, only the direct method will be considered.

The case of unit delay ($d = 1$) has been given more attention due to its simplicity. Furthermore, for the unit delay case the predictor model parameters coincide with those of the system model as seen by the following. Letting $d = 1$, the polynomials $F(z)$ and $G(z)$ in (2.5) become

$$G(z) = z^{-1}(C(z) - A(z))$$

and

$$F(z) = 1,$$

whence the predictor model (2.8) can be written as

$$C(z)\left(y^{\circ}(t|t-1)-y^{*}(t)\right)=z^{-1}\left(C(z)-A(z)\right)y(t-1)+B(z)u(t-1)-C(z)y^{*}(t)$$

$$=z^{-1}\left(1-A(z)\right)y(t-1)+B(z)u(t-1)$$

$$+z^{-1}\left(C(z)-1\right)\left(y(t-1)-y^{*}(t-1)\right)-y^{*}(t)$$

$$=\varphi^{T}(t-1)\vartheta^{\circ}-y^{*}(t)$$

where ϑ° is as in (2.14b) while $\varphi(t-1)$ is given by

$$\varphi(t-1)=[y(t-1)\cdots y(t-n),u(t-1)\cdots u(t-m-1),$$

$$(y(t-1)-y^{*}(t-1))\cdots(y(t-\ell)-y^{*}(t-\ell))]^{T}. \qquad (2.16)$$

The basic self-tuning methodology of estimation and control described above can be varied somewhat to yield certain important results which will be presented later. Following the work of [53, Kumar and Praly] the following useful definitions are made.

The Tracking Problem

An adaptive controller is to be designed that will minimize $V(t)$ in (2.2) at each instant for an arbitrary reference sequence $\{y^{*}(t)\}$ satisfying assumption (T). ∎

The Model Following Problem

An adaptive controller is to be designed that will minimize $V(t)$ at each instant for a reference sequence $\{y^{*}(t)\}$ which is asymptotically close to the output y_m of the linear model

$$H(z)y_m(t)=0$$

where

$$H(z)\overset{\Delta}{=}1-h_1z-\cdots-h_rz^r,\qquad r\geq 0,$$

in the sense that

$$\sum_{t=1}^{\infty}(y^{*}(t)-y_m(t))^2<\infty. \qquad (2.17)$$

Without loss of generality, assumptions (MF1) and (MF2) are assumed to hold. ∎ As defined above, the model following problem implies the natural assumption (T).

A simple modification of the control strategy (2.9) is noteworthy. One simply rewrites it as an orthogonality condition

$$\varphi^T(t-d)\vartheta^\circ = 0$$

where

$$\varphi(t-d) = [y(t-d)\cdots y(t-d-\bar{n}+1), u(t-d)\cdots u(t-2d-m+1), -y^*(t)\cdots-y^*(t-\ell)]^T$$

and

$$\vartheta^\circ = [g_0\cdots g_{\bar{n}-1},(bf)\cdots(bf)_{m+d-1}, 1\ c_1\cdots\cdots c_\ell]^T.$$

The dimension of these vectors has been increased by one; however, the payoff, as will be later seen, is very appealing.

To conclude this section, another variation which is particular to the model following problem adopted in [53, Kumar and Praly] is described. If $\ell > r$, consider the factorization [14, Burton]

$$C(z) = \overline{F}(z)H(z) + \overline{G}(z) \qquad (2.18a)$$

where

$$\overline{F}(z) = \sum_{i=0}^{\ell-r} \overline{f}_i z^i \qquad (2.18b)$$

and

$$\overline{G}(z) = \sum_{i=0}^{r-1} \overline{g}_i z^i. \qquad (2.18c)$$

Letting

$$\tilde{y}(t) \triangleq y_m(t) - y^*(t)$$

the term $C(z)y^*(t)$ may be written as

$$C(z)y^*(t) = C(z)\left(y_m(t) - \tilde{y}(t)\right)$$
$$= \overline{G}(z)y_m(t) - C(z)\tilde{y}(t)$$
$$= \overline{G}(z)y^*(t) - \left(C(z) - \overline{G}(z)\right)\tilde{y}(t)$$

where (2.18a) was used to obtain the second equality.
Then, $h(t) \triangleq \left(C(z) - \overline{G}(z)\right)\tilde{y}(t)$ satisfies

$$\lim_{t\to\infty} h(t) = 0 \qquad (2.19)$$

since (2.17) implies that $\sum_{t=1}^{\infty} h^2(t) < \infty$.

The asymptotic result (2.19) suggests that one replace $C(z)y^*(t)$ by $\overline{G}(z)y^*(t)$ and hence obtain a predictor model such as (2.8) with $\ell - r$ fewer parameters. Besides, if one chooses $y^*(t) = y_m(t)$ then $C(z)y^*(t) = \overline{G}(z)y^*(t)$ $\forall t$. The following choice of regression and parameter vectors is thus well motivated:

$$\varphi(t-d) = [y(t-d) \cdots y(t-d-\overline{n}+1), u(t-d) \cdots u(t-2d-m+1), -y^*(t) \cdots -y^*(t-r+1)]^T$$

and

$$\vartheta = [g_0 \cdots g_{\overline{n}-1}, (bf)_0 \cdots (bf)_{m+d-1}, \overline{g}_0 \cdots \cdots \overline{g}_{r-1}]^T.$$

2.3 Parameter Estimation and Minimum Variance Control

In Section 2.4, the presented results justify the view that the controller performs two distinct tasks. Loosely speaking, these are (i) learning and (ii) acting, and, as was mentioned in Section 2.1, are implemented via (i) parameter estimation and (ii) MV control. In this section, parameter estimation will first be considered followed by a discussion of the concurrent operation of optimal (MV) control and parameter estimation. The various assumptions used are listed in Appendix A for easy reference while the various regression and parameter vectors are listed in Appendix B.

2.3.1 Parameter Estimation: Principal Algorithms and their Properties

Two main types of estimation schemes are examined, namely, the stochastic approximation type and the least squares type. They are by far the most commonly employed schemes in algorithms found in the current literature of stochastic adaptive control. Informally, these parameter estimation algorithms can be thought of as stochastic versions of the well known numerical methods of steepest descent and the Newton method [56, Ljung and Söderström], [43, Goodwin and Sin]. The stochastic approximation type algorithms are presented first and it is felt that some introductory remarks are in order.

Stochastic Approximation Algorithms

The origin of these algorithms is attributed to the early paper [62, Robbins and Monro] where a recursive scheme for the unique solution ϑ^* of

$$M(\vartheta^*) = \mathrm{E}[Y(\omega; \vartheta^*)] = \alpha \qquad (2.20)$$

is proposed. An experiment parameterized by ϑ is performed and an observation $y(\vartheta)$ of $Y(\omega; \vartheta)$ is recorded. As an example, one can think of recording noisy measurements at the output of a linear ϑ-parameterized system where this output is expected to vary about some known value α. It is assumed however that the exact nature of $M(\cdot)$ or that of the distribution $H_Y(\cdot)$ of Y is not known, but that they satisfy certain technical assumptions. Then, the recursion

$$\vartheta(t+1) - \vartheta(t) = a(t)\left(\alpha - y(\vartheta(t))\right) \qquad (2.21)$$

produces a sequence $\{\vartheta(t)\}$ which converges to ϑ^* in a mean square sense, i.e.

$$\lim_{t\to\infty} \mathrm{E}[(\vartheta(t) - \vartheta^*)^2] = 0.$$

It is remarked that the sequence of gains $\{a(t)\}$ plays the key role in the convergence of $\vartheta(t)$. While $\lim a(t) = 0$ is a necessary condition for sample path convergence, $a(t)$ cannot go to zero very rapidly else it would then be impossible to "erase", on the average, the effects of errors. On the other hand, the effect of new noisy measurements on extensive experience should not be allowed to alter the latest estimate substantively. These statements are reflected respectively by the conditions required in [62, Robbins and Monro], namely,

$$\sum_{t=1}^{\infty} a(t) = \infty \qquad \text{and} \qquad \sum_{t=1}^{\infty} a^2(t) < \infty.$$

The recursion scheme (2.21) was named stochastic approximation in [62, Robbins and Monro] and results such as convergence w.p.1, were subsequently obtained by [51, Dvoretzky].

Consider now the problem at hand, namely, the estimation of ϑ°, the parameter vector of the predictor model. When the system is under adaptive control as given by (2.9),

$$y(t) = \varphi^T(t-1)\vartheta^\circ + e(t)$$

where

$$e(t) = y(t) - y^*(t).$$

Lacking knowledge of ϑ°, an estimate $\hat{\vartheta}$ of ϑ° is sought via the minimization over ϑ of

$$V(\vartheta) = \mathrm{E}[(y(t) - \varphi^T(t-1)\vartheta)^2].$$

It then follows that

$$\left.\frac{dV(\vartheta)}{d\vartheta}\right|_{\vartheta=\hat{\vartheta}} = 0. \tag{2.22}$$

Differentiating inside the expectation and defining

$$Q(\vartheta) \triangleq -\varphi(t-1)(y(t) - \varphi^T(t-1)\vartheta), \tag{2.23}$$

(2.22) translates to

$$\mathrm{E}[Q(\hat{\vartheta})] = 0.$$

This is similar to (2.20) where Q corresponds to Y and $\alpha = 0$. Following the idea of Robbins and Monro one then gets

$$\begin{aligned}
\hat{\vartheta}(t+1) &= \hat{\vartheta}(t) - a(t)Q\left(\hat{\vartheta}(t)\right) \\
&= \hat{\vartheta}(t) + a(t)\varphi(t)\left(y(t+1) - \varphi^T(t)\hat{\vartheta}(t)\right)
\end{aligned} \tag{2.24}$$

where $a(t)$ is as in (2.21). Equation (2.24) has the form of the so-called stochastic gradient algorithm which will be presented shortly. The name gradient probably comes from descent techniques in unconstrained optimization (see e.g. [36, Jacoby et.al.]). There, one finds iterations having the general form

$$\hat{\vartheta}(t+1) = \hat{\vartheta}(t) + \lambda(t)s(t) \tag{2.25}$$

where $s(t)$ is a descent direction and $\lambda(t)$ a descent step-length. Clearly the direction of steepest descent is that of the negative gradient of the function in question. In the present case,

$$s(t) = \mathrm{E}[-Q(\hat{\vartheta}(t))].$$

By dropping the expectation (2.25) becomes

$$\hat{\vartheta}(t+1) = \hat{\vartheta}(t) + \lambda(t)\varphi(t)(y(t+1) - \varphi^T(t)\hat{\vartheta}(t)) \tag{2.26}$$

which has the interpretation that on the average adjustments are made in the directions of $s(t)$. With this motivation, the stochastic gradient algorithm in its most commonly known form is presented next.

The Stochastic Gradient Algorithm (SG)

For $t \geq 1$, $d \geq 1$,

$$\hat{\vartheta}(t) = \hat{\vartheta}(t-d) + \frac{1}{r(t-d)}\varphi(t-d)\left(y(t) - \varphi^T(t-d)\hat{\vartheta}(t-d)\right), \qquad (2.27a)$$

$$r(t-d) = r(t-d-1) + \varphi^T(t-d)\varphi(t-d), \qquad (2.27b)$$

where $r(1-d) = 1$ and $\hat{\vartheta}(1-d), \cdots\cdots, \hat{\vartheta}(0)$ are arbitrary. ∎

Attention now is drawn to the fact that, when the system (2.1) is under the control specified in (2.15), the regression vector $\varphi(\cdot)$ contains a priori predictions $y^*(\cdot)$ as in (2.8b) or a priori prediction errors $y(\cdot) - y^*(\cdot)$ as in (2.16). To see this, notice that when computing $u(t)$ from (2.15), $\hat{\vartheta}(t)$ is available but $\varphi(t)$ contains $y^*(t+d-1), \cdots, y^*(t + d - \ell)$ which are equal to $\varphi(t-1)\hat{\vartheta}(t-1), \cdots, \varphi(t-\ell)\hat{\vartheta}(t-\ell)$ via (2.15). This means that full use of the latest available information is not being made, and in this vein one defines the a posteriori predictions as

$$\bar{y}(t) \triangleq \varphi(t-d)\hat{\vartheta}(t-d+1)$$

where

$$\varphi(t-d) = [y(t-d)\cdots y(t-d-\bar{n}+1), u(t-d)\cdots u(t-2d-m+1), -\bar{y}(t-1)\cdots -\bar{y}(t-\ell)]^T$$

replaces (2.8b), or replacing (2.16),

$$\varphi(t-1) = [y(t-1)\cdots y(t-n), u(t-1)\cdots u(t-m-1),$$
$$(y(t-1) - \bar{y}(t-1))\cdots(y(t-\ell) - \bar{y}(t-\ell))]^T.$$

The stochastic gradient algorithm (2.27) with the above φ's was introduced in [23, Chen] where the name *Quasi-Least Squares Algorithm* was coined. The motivation was obtained from the algorithm of [65, Sin and Goodwin] who made use of a posteriori predictions. At this point it should be mentioned that every effort has been made to present the various algorithms and their modifications under their most widely used name. However, the reader is warned that there is no completely standard usage. For example, in [43, Goodwin and Sin], the Quasi-Least Squares Algorithm above will be found under the name Stochastic Gradient Algorithm.

Least Squares Algorithms

The simplicity of the above algorithms comes at the expense of slow convergence as will be seen in the next chapter. An algorithm with better convergence properties is the well known least squares algorithm which results from the minimization over ϑ of

$$\overline{V}_N(\vartheta) = \sum_{t=1}^{N} (y(t) - \varphi^T(t - d)\vartheta)^2.$$

Alternatively (see e.g. [56, Ljung and Söderström], [43, Goodwin and Sin]), it can also be viewed as a search for the zeros of $dV/d\vartheta$ using the Newton method whereby

$$\hat{\vartheta}(t) = \hat{\vartheta}(t - 1) + \left[\frac{d^2V}{d\vartheta^2}\Big|_{\hat{\vartheta}(t-1)}\right]^{-1} \frac{dV}{d\vartheta}\Big|_{\hat{\vartheta}(t-1)}.$$

Arguing heuristically, in a way similar to Robbins and Monro, one can approximate the above via

$$\hat{\vartheta}(t) = \hat{\vartheta}(t - 1) + \mathbf{P}^{-1}(t)\left[-\varphi(t - 1)\left(y(t) - \varphi^T(t - 1)\hat{\vartheta}(t - 1)\right)\right] \qquad (2.28)$$

where

$$\mathbf{P}(t) = \frac{d^2V}{d\vartheta^2}\Big|_{\hat{\vartheta}(t-1)} = \mathrm{E}\left[\varphi(t)\varphi^T(t)\right]$$

can be recursively computed using the Robbins-Monro algorithm to solve

$$\mathrm{E}\left[\varphi(t)\varphi^T(t) - \mathbf{P}(t)\right] = 0.$$

This gives

$$\mathbf{P}(t) = \mathbf{P}(t - 1) + \gamma(t - 1)\left[\varphi(t)\varphi^T(t) - \mathbf{P}(t - 1)\right] \qquad (2.29)$$

where $\{\gamma(t)\}$ is some sequence of gains. Choose in particular the harmonic sequence $\gamma(t) = 1/t$, which is square summable but not summable, and let $\mathbf{R}(t) = \mathbf{P}^{-1}(t)/t$. Using the well known Matrix Inversion Lemma (see Appendix C), equations (2.28) and (2.29) give rise to the celebrated recursive least squares (RLS) algorithm

$$\hat{\vartheta}(t) = \hat{\vartheta}(t - 1) + \mathbf{R}(t - 1)\varphi(t - 1)\left(y(t) - \varphi^T(t - 1)\hat{\vartheta}(t - 1)\right) \qquad (2.30a)$$

$$\mathbf{R}(t - 1) = \mathbf{R}(t - 2) - \frac{\mathbf{R}(t - 2)\varphi(t - 1)\varphi^T(t - 1)\mathbf{R}(t - 2)}{1 + \varphi^T(t - 1)\mathbf{R}(t - 2)\varphi(t - 1)} \qquad (2.30b)$$

In fact, these equations are exact and can be deduced from a standard minimization of $\overline{V}_N$ (see e.g. [15, Caines]). The direct numerical implementation of the RLS algorithm is numerically unstable, and even with stable implementations, updating $\mathbf{R}(t)$ is an ill-conditioned problem. Thus, modifications of the least squares algorithm have been developed whereby the condition number of the gain matrix $\mathbf{R}(\cdot)$ is checked at each iteration. While these modifications augment the computational requirements, they make possible certain convergence results. Such a modification was first introduced in [65, Sin and Goodwin] where the name *Modified Least Squares Algorithm* was coined. Shortly after, [27, Chen] gave a different modification and proved stronger results. The aforementioned algorithms (which use a posteriori regressions) are as follows.

The Modified Least Squares Algorithm (MLS)

For $t \geq 1$, $d \geq 1$,

$$\widehat{\vartheta}(t) = \widehat{\vartheta}(t-d) + a(t-d)\mathbf{R}(t-d)\varphi(t-d)\left(y(t) - \varphi^T(t-d)\widehat{\vartheta}(t-d)\right), \quad (2.31a)$$

$$\mathbf{R}'(t-d) = \mathbf{R}(t-2d) - \frac{\mathbf{R}(t-2d)\varphi(t-d)\varphi^T(t-d)\mathbf{R}(t-2d)}{1 + \varphi^T(t-d)\mathbf{R}(t-2d)\varphi(t-d)} \quad (2.31b)$$

where $\widehat{\vartheta}(1-d), \cdots, \widehat{\vartheta}(0)$ are arbitrary, $\mathbf{R}(1-2d) > 0$, and $\mathbf{R}(t)$ is given by either Modification 1 or Modification 2 below.

Modification 1 [27, Chen]

Let

$$r(t-d) = r(t-d-1) + \varphi^T(t-d)\varphi(t-d), \qquad r(-1) = 1,$$

and $\lambda_{\max}(t-d)$ and $\lambda_{\min}(t-d)$ the maximum and minimum eigenvalues of $\mathbf{R}'(t-d)$ respectively. If

$$\frac{\lambda_{\max}(t-d)}{\lambda_{\min}(t-d)} \leq K_1 \qquad \text{and} \qquad \varphi^T(t-d)\mathbf{R}'(t-d)\varphi(t-d) \leq K_2 < 1$$

then

$$\mathbf{R}(t-d) = \mathbf{R}'(t-d) \qquad \text{and} \qquad a(t-d) = 1$$

else

$$\mathbf{R}(t-d) = \frac{r(t-d-1)}{r(t-d)}\mathbf{R}'(t-d) \quad \text{and} \quad a(t-d) = \frac{1}{1 + \varphi^T(t-d)\mathbf{R}(t-d)\varphi(t-d)}.$$

Modification 2 [65, Sin and Goodwin]

Let

$$r(t - d) = r(t - d - 1)\left(1 + \varphi^T(t - d)\mathbf{R}(t - 2d)\varphi(t - d)\right) \qquad r(-1) > 0,$$

$\lambda_{\max}(t - d)$ the maximum eigenvalue of $\mathbf{R}'(t - d)$, and $a(t) = 1$. If

$$r(t - d)\lambda_{\max}(t - d) \le K, \qquad 0 < K < \infty,$$

then

$$\mathbf{R}(t - d) = \mathbf{R}'(t - d)$$

else

$$\mathbf{R}(t - d) = \frac{K}{r(t - d)\lambda_{\max}(t - d)}\mathbf{R}'(t - d).$$

∎

The algorithms presented so far are only a select few of many possible ones (see e.g. [56, Ljung and Söderström]), and detached from their use in a feedback control scheme, they enjoy certain convergence properties which are outlined in the theorems that follow. Prior to their presentation, a remark concerning the quantity $r(t)$ is in order. The set of events $\Gamma = \{\omega : \lim_{t\to\infty} r(t) < \infty\}$ is rare and uninteresting because it implies that $\lim_{t\to\infty} y(t) = 0$ and $\lim_{t\to\infty} u(t) = 0$ and in some cases, that $\lim_{t\to\infty} w(t) = 0$ [23, 25, Chen]. Consequently, in the following results, it will be assumed that $\omega \in \Omega\backslash\Gamma$ where $\Omega\backslash\Gamma = \{\omega : \lim_{t\to\infty} r(t) = \infty\}$. In the applications of these results in adaptive control it is proved that either $P(\Omega\backslash\Gamma) = 1$ or the results are established on both Γ and $\Omega\backslash\Gamma$ and hence on all of Ω (see e.g. Corrigendum in [41, Goodwin et.al.]). Let

$$\Phi(t + 1, i) = \begin{cases} \left(I - r^{-1}(t)\varphi(t)\varphi^T(t)\right)\Phi(t, i) & t > i \\ I & t = i \end{cases}$$

where $i \ge 0$ and, as before, $r(t) = 1 + \sum_{i=1}^{t}\|\varphi(i)\|^2$.

Theorem 2.1 [31, Chen and Guo] (SG)

Consider algorithm (2.27) with (B2.1) where $n' = n, m' = m$, and $\ell' = \ell$. Let assumptions (W1) and (W3) hold. Then, if $\ell = 0$,

$$\lim_{t\to\infty} \widehat{\vartheta}(t) = \vartheta^\circ \quad \text{a.s.} \quad \text{iff} \quad \lim_{t\to\infty} \Phi(t, 0) = 0 \quad \text{a.s.}$$

If $\ell \neq 0$ and (M6) also holds,

$$\lim_{t \to \infty} \widehat{\boldsymbol{\vartheta}}(t) = \boldsymbol{\vartheta}^\circ \quad \text{a.s. if} \quad \lim_{t \to \infty} \Phi(t,0) = 0 \quad \text{a.s.}$$

∎

The sufficient condition for the convergence of $\widehat{\boldsymbol{\vartheta}}(\cdot)$ is not verifiable a priori and is therefore of limited practical interest. This is also the case with the conditions

$$(i) \qquad \lim_{t \to \infty} \frac{1}{t} \sum_{i=1}^{t} \varphi(i)\varphi^T(i) = \mathbf{R} > 0 \quad \text{a.s.} \quad \mathbf{R} \in \mathfrak{R}^{\rho \times \rho}, \varphi \in \mathfrak{R}^\rho, \qquad (2.32)$$

$$(ii) \qquad \mathbf{R}(t) = \sum_{i=1}^{t} \varphi(i)\varphi^T(i) + \frac{1}{\rho}I, \quad \mathbf{R}(\cdot) \in \mathfrak{R}^{\rho \times \rho}, \varphi(\cdot) \in \mathfrak{R}^\rho,$$

$$\text{and} \qquad \frac{\lambda_{\max}}{\lambda_{\min}}(\mathbf{R}(t)) \leq \gamma(\omega), \qquad (2.33)$$

and Chen's condition (G1),

$$(iii) \qquad \exists T(\omega) > 0, \alpha(\omega) > 0, \quad \text{and} \quad \beta(\omega) > 0 \quad \text{s.t.}$$

$$\sum_{i=m(t)}^{m(t+a)} \frac{\varphi(i)\varphi^T(i)}{r(i)} \geq \beta I \quad \forall t \geq T, \quad \forall \omega \in \{\omega : \lim_{t \to \infty} r(t) = \infty\}$$

$$\text{where} \quad m(t) = \max\{n : p(n) \leq t\}, \; t \geq 0, \; p(n) = \sum_{i=0}^{n-1} \|\varphi(i)\|^2/r(i),$$

which have appeared so far as sufficient conditions for the convergence of the parameter estimation schemes presented here. As the following complementary result shows, the sufficient condition of theorem 2.1 is the weakest among them.

Corollary 2.1a [31, Chen and Guo]

For any fixed ω, (2.32) $\Rightarrow$ (2.33) $\Rightarrow$ (G1) $\Rightarrow \lim_{t \to \infty} \Phi(t,0) = 0$. ∎

Further to this, perhaps the only result so far on the rate of convergence of algorithm (2.27) has been given by

Corollary 2.1b [32, Chen and Guo]

If

$$\limsup_{t \to \infty} \frac{r(t)}{r(t-1)} < \infty$$

and $\exists N_0(\omega), M(\omega)$ such that

$$\frac{\lambda_{\max}}{\lambda_{\min}}(\mathbf{R}(t)) \leq M(\log r(t))^{\frac{1}{4}} \quad \text{a.s.} \quad \forall t \geq N_0,$$

then

$$\lim_{t \to \infty} \Phi(t, 0) = 0.$$

If in addition $\ell' = 0$,

$$\|\widehat{\boldsymbol{\vartheta}}(t) - \boldsymbol{\vartheta}^\circ\| = o\left(\log^{-\delta_2} r(t)\right) \qquad \delta_2 > 0.$$

If, more restrictively,

$$\frac{\lambda_{\max}}{\lambda_{\min}}(\mathbf{R}(t)) \le \gamma(\omega) < \infty \qquad \text{a.s. with} \qquad \ell' = 0,$$

then

$$\|\widehat{\boldsymbol{\vartheta}}(t) - \boldsymbol{\vartheta}^\circ\| = o\left(r(t)^{-\delta_1}\right), \, \delta_1 \in \left(0, \frac{1}{\dim \boldsymbol{\vartheta}}\right).$$

∎

It is interesting to note that the above results show that parameter convergence can occur even when $\mathbf{R}(t)$ is not well conditioned. A convergence result for the QLS algorithm is presented next.

Theorem 2.2 [23, 25, Chen] (QLS)

Consider algorithm (2.27) with (B1.2) where $d = 1$. Let assumptions (M2), (M6), (W1), (W9), (G1), and (G2) hold, in addition to

$$\lim_{t \to \infty} \frac{r(t)}{r(t-1)} \le \gamma, \qquad \forall t \ge 1.$$

Then

$$\lim_{t \to \infty} \widehat{\boldsymbol{\vartheta}}(t) = \boldsymbol{\vartheta}^\circ \qquad \text{a.s.}$$

∎

For the LS and MLS algorithms the following results are very general.

Theorem 2.3 [26, Chen] (LS)

Consider algorithm (2.30) with $\mathbf{R}(-1) > 0$, (B1.2) and $d = 1$. In addition to

$$\frac{\lambda_{\max}}{\lambda_{\min}}(\mathbf{R}(t)) \le K_1, \qquad \forall t \ge 1, \tag{2.34}$$

let assumptions (M2), (M4), (W1), (W7), (W9), and (G2) hold. Then

$$\lim_{t \to \infty} \widehat{\boldsymbol{\vartheta}}(t) = \boldsymbol{\vartheta}^\circ \qquad \text{a.s.} \tag{2.35}$$

and

$$\|\widehat{\boldsymbol{\vartheta}}(t) - \boldsymbol{\vartheta}^\circ\| = o\left(r^{\delta-\frac{1}{2}}(t)\right) \quad \text{a.s.} \quad \forall \delta \in (\frac{\epsilon}{2}, \frac{1}{2}], \quad \epsilon \in (0,1). \tag{2.36}$$

For $\ell = 0$, if (2.34) is replaced by

$$\exists\, K_2(\omega) \quad \text{and} \quad \alpha \in [0, \frac{1}{2}) \quad \text{s.t.} \quad \lambda_{\max}(\mathbf{R}(t)) \leq \frac{K_2(\omega)}{r^{1-\alpha}(t)},$$

and (W9) is omitted then (2.35) holds with (2.36) replaced by

$$\|\widehat{\boldsymbol{\vartheta}}(t) - \boldsymbol{\vartheta}^\circ\| = o\left(r^{\delta-\frac{1}{2}}(t)\right) \quad \text{a.s.} \quad \forall \delta \in (\alpha + \frac{\epsilon}{2}, \frac{1}{2}], \quad \epsilon \in (0,1).$$

■

Theorem 2.4 [27, Chen] (MLS)

Consider algorithm (2.31) with Modification 1, (B2.2), $d = 1$, and $\mathbf{R}(-1) > 0$. Let assumptions (M2), (M4), (W1), (W3), and (G1) hold. Then,

$$\lim_{t\to\infty} \widehat{\boldsymbol{\vartheta}}(t) = \boldsymbol{\vartheta}^\circ \quad \text{a.s.}$$

■

2.3.2 Remarks on Parameter Estimation under Minimum Variance Control

It was shown in Section 2.2 that the control $u(t)$ which minimizes (2.2) is given as the solution of

$$\boldsymbol{\varphi}^T(t)\boldsymbol{\vartheta}^\circ = y^*(t + d)$$

or by

$$\boldsymbol{\varphi}(t)\boldsymbol{\vartheta}^\circ = 0$$

depending on the choice of regression and parameter vectors, where $\boldsymbol{\vartheta}^\circ$ contains the actual predictor model parameters. The self-tuning philosophy suggests that in the case of unknown parameters their latest estimate be used instead, i.e. to compute $u(t)$ via

$$\boldsymbol{\varphi}^T(t)\widehat{\boldsymbol{\vartheta}}(t) = y^*(t + d) \tag{2.37a}$$

or via

$$\boldsymbol{\varphi}^T(t)\widehat{\boldsymbol{\vartheta}}(t) = 0. \tag{2.37b}$$

As will be seen in what follows, the above choice of control is equivalent, in a sense to be made precise later, to knowing the parameters with certainty. This property of certainty equivalence is intimately connected with the information I that the estimator sends to the controller (see Fig. 2.1). Formal but very readable discussions on certainty equivalence and related concepts can be found in [8, Bar-Shalom and Tse] and [61, Patchell and Jacobs].

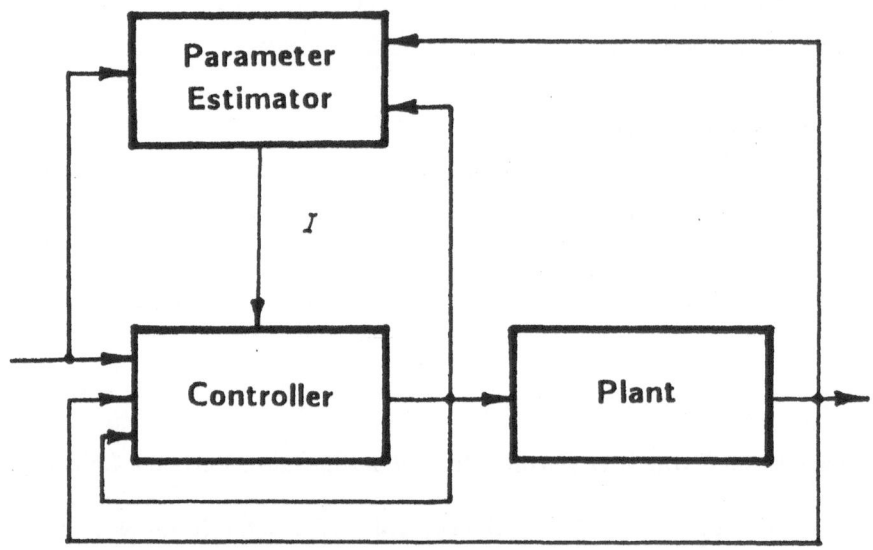

Figure 2.1 A Parameter Adaptive Control Scheme

Loosely speaking, it is intuitively clear that when simultaneous consistent estimation and optimal tracking is the objective, the control generated by the feedback scheme (2.37) will not necessarily probe all independent subsystems which comprise the actual system. However, (2.37) *will* result in a sufficiently probing control action if, for example, (2.32) is satisfied. This condition is known in the literature as a persistence of excitation condition. Although (2.32) as well as weaker conditions can guarantee consistency of the estimated parameters, they do not hold in general. For example, (2.32) is not necessarily satisfied when (2.37) is used and so a disturbance or dither is added to yield what has been most appropriately named in [16, Caines] continually disturbed controls

(CDC). The control $u(t)$ is then determined via

$$\varphi^T(t)\widehat{\vartheta}(t) = y^*(t+d) + v(t) \qquad (2.37c)$$

where $\{v(t)\}$ is a white noise, zero mean scalar process. Although such a strategy results in a sub-optimal performance, it has the definite advantage of being easily implementable. In practice, such a sub-optimal performance might be tolerated. Further understanding will be gained if the sum $y^*(t+d) + \varepsilon(t)$ is interpreted as the reference sequence itself. It is then appealing to think of this new reference sequence as "sufficiently rich" to result in a control action which is continually disturbed. The idea of sufficient richness has been formalized in [13, Boyd and Sastry] for a deterministic adaptive control problem while [53, Kumar and Praly] makes the following analogous definition for the stochastic case.

Definition 2.1

A scalar sequence $\{y^*(t)\}$ is said to be strongly sufficiently rich of order ℓ if ℓ is the largest non-negative integer for which there exist integers n and T and a real $\varepsilon > 0$ such that

$$\sum_{k=t+1}^{t+n} [y^*(k-1)\cdots y^*(k-\ell)]^T[y^*(k-1)\cdots y^*(k-\ell)] \geq \varepsilon I_\ell, \qquad \forall t > T, \qquad (2.38)$$

where I_ℓ is the identity matrix of order ℓ. If further (2.38) holds for all ℓ, $\{y^*(t)\}$ is be said to be strongly sufficiently rich of arbitrarily large order. ∎

For example, it is easy to verify that a constant sequence is strongly sufficiently rich of order 1. On the other hand, the reference sequence specified by the r.h.s. of (2.37c) is strongly sufficiently rich of arbitrarily large order. For the model following problem, the following lemma of [53, Kumar and Praly] establishes a useful result.

Lemma 2.1

In the model following problem, $\{y_m(t)\}$ and $\{y^*(t)\}$ are strongly sufficiently rich of order r. ∎

Although a connection between persistency of excitation as in (2.32) and sufficient richness of $\{y^*(t)\}$ as in (2.38) no doubt exists, its precise nature has not yet been well exposed in the stochastic case. For some work on the deterministic case see [1, Anderson and Johnson].

2.4 Principal Results on Self-Tuning Control and System Identification

The theorems presented in this section summarize the most recent results on the problem of adaptive tracking and consistent parameter estimation. It is important to note that in the tracking problem, the minimization of (2.2) is of course subject to a stability requirement on the closed loop system. Consequently, in addition to some kind of convergence of the tracking error, the inputs and outputs are required to be bounded in the sense that

$$\limsup_{N\to\infty}\frac{1}{N}\sum_{t=1}^{N}u^2(t) < \infty \qquad \text{a.s.}$$

$$\limsup_{N\to\infty}\frac{1}{N}\sum_{t=1}^{N}y^2(t) < \infty \qquad \text{a.s.}$$

Such a global convergence result was first proved in [41, Goodwin et.al.] for the unit delay coloured noise case and subsequently, [44, Goodwin et.al.] proved it for the general delay case. The following theorem due to [41, Goodwin, Ramadge and Caines] states the important conclusions.

Theorem 2.5 [41,44, Goodwin et. al.] (SG)

Consider the tracking problem for the system (2.1) using algorithm (2.27) with (B1.1) and the adaptive control (2.37a). Let assumptions (M1), (M3), (M5), (W1), (W2), (W5) and (W8) hold. Then

$$\limsup_{N\to\infty}\frac{1}{N}\sum_{t=1}^{N}y^2(t) < \infty \qquad \text{a.s.} \tag{2.39a}$$

$$\limsup_{N\to\infty}\frac{1}{N}\sum_{t=1}^{N}u^2(t) < \infty \qquad \text{a.s.} \tag{2.39b}$$

and

$$\lim_{N\to\infty}\frac{1}{N}\sum_{t=1}^{N}E[(y(t)-y^*(t))^2|\mathcal{F}_{t-d}] = \gamma^2 \qquad \text{a.s.} \tag{2.39c}$$

where γ^2 is given by (2.7). ∎

It is noteworthy that while [44, Goodwin, et.al.] does not discuss at all the possibility of division by zero in the computation of the control, [41, Goodwin, et.al.] leaves it to the reader to inductively verify that such an event has zero probability. This was subsequently proved in a general case by [57, Meyn and Caines] and their condition is displayed in (W8).

It is evident from the discussion in Section 2 that the result expressed in (2.39c) cannot be improved upon even if the true parameters were known. In this asymptotic sense, the control strategy (2.37a) is certainty equivalent. While the above result deals with the SG algorithm, identical results hold for the QLS and MLS algorithms as expressed in the following theorems.

Theorem 2.6 [43, Goodwin and Sin] (QLS)

Consider the tracking problem for the system (2.1) using algorithm (2.27) with (B1.2) and the adaptive control (2.37a). Let assumptions (M1), (M3), (M6), (W1), (W2), (W5) and (W8) hold. Then the results (2.39) of theorem 2.5 hold.　■

Theorem 2.7 [65, Sin and Goodwin] (MLS)

Consider the tracking problem for the system (2.1) with $d = 1$, algorithm (2.31) with Modification 2 and (B1.2) and under the adaptive control (2.37a). Let assumptions (M1), (M3), (M4), (W1), (W2), (W5) and (W8) hold. Then the results (2.39) of theorem 2.5 hold.　■

Corollary 2.7a [27, Chen]

If in theorem 2.7, Modification 1 with (B2.2) is used instead, and if assumptions (W2) and (W5) are respectively replaced by (W3) and (W6), then (2.39a) and (2.39b) hold, while (2.39c) is replaced by

$$\lim_{N \to \infty} \frac{1}{N} \sum_{t=1}^{N} (y(t) - y^*(t))^2 = \sigma_w^2 \qquad (2.40)$$

■

Remark: The slightly strengthened conclusion (2.40) also applies to theorems 2.5 and 2.6 upon replacement of (W5) by (W7).　■

The theorems presented so far establish the desired results of stability and asymptotic cost optimality. Recently, the scope of these objectives has been expanded to include the simultaneous consistency of the parameter estimates. Initial steps in this direction were taken by [54, Lafortune] and [20, Caines and Lafortune], which were subsequently followed by numerous papers on the topic. From the discussion in section

2.2, it is clear that if the parameter estimates converge to yield some optimal control law, asymptotic cost optimality will then be achieved. In this vein, define

$$D = \{\boldsymbol{\vartheta} : \boldsymbol{\vartheta} \text{ produces a minimum variance control law}\}.$$

For example, this set contains multiples of the true predictor model parameter vector $\boldsymbol{\vartheta}^\circ$ or vectors for which appropriate cancellations take place to give an optimal control law. Hence, convergence to an element of D does not necessarily imply convergence to the true controller parameters. The following result is relevant.

Theorem 2.8 [9, Becker et.al.] (SG)

Consider the model following problem for the system (2.1) with $d = 1$. Let $y^*(t) \equiv 0$ and suppose assumptions (M1), (M3), (W1), (W2), (W5), (W7), and (W8) hold. Using (B3) with algorithm (2.27) and (M5), or algorithm (2.31) with either of its modifications and (M4), and under adaptive control (2.37b) gives

$$\lim_{N \to \infty} \frac{1}{N} \sum_{t=1}^{N} \mathbb{1}_O\left(\widehat{\boldsymbol{\vartheta}}(t)\right) = 1$$

for almost all ω and for every open set $O \subseteq D$. ∎

In other words, the theorem above says that, for an open set O as "close" to D as one likes, $\widehat{\boldsymbol{\vartheta}}(t) \in O$ infinitely often. Further to this result, the following theorem establishes the almost sure convergence of the parameter estimates of the SG algorithm to a multiple of the true parameter vector $\boldsymbol{\vartheta}^\circ$.

Theorem 2.9 [9, Becker et.al.] (SG)

Consider the situation of theorem 2.8 using algorithm (2.27) with (M5). Further assume that (M7) holds. Then, with $k(\omega)$ an a.s. constant random variable,

$$\lim_{t \to \infty} \widehat{\boldsymbol{\vartheta}}(t) = k(\omega)\boldsymbol{\vartheta}^\circ \qquad \text{a.s.}$$

In addition, if $\|\widehat{\boldsymbol{\vartheta}}(0)\| > \|\boldsymbol{\vartheta}^\circ\|$,

$$\lim_{t \to \infty} \widehat{\boldsymbol{\vartheta}}(t) \neq \boldsymbol{\vartheta}^\circ \qquad \text{a.s.} \tag{2.41}$$

while if $\varphi(0) \neq 0$,

$$\text{Prob} \{\omega : \lim_{t \to \infty} \widehat{\boldsymbol{\vartheta}}(t) \neq \boldsymbol{\vartheta}^\circ\} > 0.$$

The results of the two previous theorems do not invoke the ideas of persistence of excitation, sufficient richness, or CDC. In fact, in the framework of the said theorems, persistence of excitation holds $w.p.$ 0 and moreover, (2.41) shows that in general, the parameter estimates will not converge to ϑ°. The sufficient richness idea is exploited in the following theorem.

Theorem 2.10 [53, Kumar and Praly] (SG)

Consider the system (2.1) with $d = 1$, under adaptive control (2.37b) and using algorithm (2.27). Let assumptions (M2), (M3), (M5), (M8), (W1), (W2), (W7) and (W8) hold. Then

(i)   for the tracking problem with (B4), if $\{y^*(t)\}$ is strongly sufficiently rich of order $q > m + \ell$,

(1) $$\lim_{t\to\infty} \widehat{\vartheta}(t) = \varsigma(\omega)\vartheta^\circ \quad \text{a.s. where} \quad \varsigma(\omega) \neq 0, \ \varsigma(\omega) < \infty \qquad \text{a.s.} \qquad (2.42)$$

(2) $$\lim_{t\to\infty} \frac{1}{\widehat{\gamma}_0(t)}[\widehat{g}_0(t) - \widehat{c}_1(t)\cdots\widehat{g}_{n-1}(t) - \widehat{c}_n(t), (\widehat{bf})_0(t)\cdots(\widehat{bf})_m(t),$$
$$\widehat{c}_1(t)\cdots\widehat{c}_\ell(t)] = [-a_1\cdots - a_n, b_0\cdots b_m, c_1\cdots c_\ell] \qquad \text{a.s.} \qquad (2.43)$$

where $\widehat{c}_i(t) = 0 \quad \text{for} \quad i > \ell$

(3) $$\lim_{t\to\infty} \frac{1}{(\widehat{bf})_0(t)}[\widehat{g}_0(t)\cdots\widehat{g}_{\overline{n}-1}(t), (\widehat{bf})_1(t)\cdots(\widehat{bf})_m(t), \widehat{\gamma}_0(t)\widehat{c}_1(t)\cdots\widehat{c}_\ell(t)] =$$
$$\frac{1}{b_0}[c_1 - a_1\cdots c_{\overline{n}} - a_{\overline{n}}, b_1\cdots b_m, 1 \ c_1\cdots c_\ell] \qquad \text{a.s.} \qquad (2.44)$$

where $a_i = 0 \quad i > n, \qquad c_i = 0 \quad i > \ell.$

(ii)   For the model following problem with (B3) or (B4), (2.42) holds. Further, for $r > \ell$ and using (B4), (2.43) and (2.44) hold while if $r \leq \ell$, using (B3),

$$\lim_{t\to\infty} \frac{1}{(\widehat{bf})_0(t)}[\widehat{g}_0(t)\cdots\widehat{g}_{\overline{n}-1}(t), (\widehat{bf})_1(t)\cdots(\widehat{bf})_m(t), \widehat{g}_0(t)\cdots\widehat{g}_{r-1}(t)] =$$
$$\frac{1}{b_0}[c_1 - a_1\cdots c_{\overline{n}} - a_{\overline{n}}, b_1\cdots b_m, \overline{g}_0\cdots\overline{g}_{r-1}] \qquad (2.45)$$

where $a_i = 0 \quad i > n, \ c_i = 0 \quad i > \ell$ with $\overline{g}_i$ given in (2.18). ∎

Equations (2.44) and (2.45) show that the adaptive controller is actually self-tuning i.e. its parameters converge to the true parameters of the MV controller. Also evident is the fact that the number of parameters which must be estimated in order to optimally reject the coloured noise is related to the richness of the reference sequence. A similar result was independently reported in [32, Chen and Guo]. Their recent study proves global convergence and consistent parameter estimation for the SG algorithm (2.27) using (B2.1) where the reference sequence $\{y^*(t)\}$ is subject to

$$\lim_{N \to \infty} \inf \lambda_{\min} \left[\frac{\log^{\frac{1}{4}} N}{N} \sum_{t=1}^{N} Y^*(t) Y^{*T}(t) \right] \neq 0$$

where

$$Y^*(t) = [y^*(t) \cdots y^*(t - n - m + 1)]^T.$$

The results of the previous theorem suggest that one might artificially alter the order of strong sufficient richness of $\{y^*(t)\}$ to achieve parameter convergence. This is related to the idea of CDC. This latter approach has been studied for the MLS and SG algorithms and the results are as follows.

Theorem 2.11 [29, Chen and Caines] (SG)

Consider the tracking problem for system (2.1) with $d = 1$. Let assumptions (M2), (M3), (M5), (M8), (W1), (W3) and (W4) hold and assume that algorithm (2.27) with (B2.1) is used in conjunction with the adaptive control (2.37c). Then (2.39a) and (2.39b) hold together with

$$\lim_{N \to \infty} \frac{1}{N} \sum_{t=1}^{N} (y(t) - y^*(t))^2 = \sigma_w^2 + \sigma_v^2 \qquad \text{a.s.} \tag{2.46}$$

and

$$\lim_{t \to \infty} \widehat{\vartheta}(t) = \vartheta^\circ \qquad \text{a.s.} \tag{2.47}$$

■

Theorem 2.12 [27, Chen] (MLS)

Consider the tracking problem for the system (2.1) with $d = 1$. Let assumptions (M2), (M3), (M4), (M8), (W1), (W3), and (W4) hold and assume that algorithm (2.31) with Modification 1 and (B2.2) is used in conjunction with the adaptive control (2.37c). Then (2.39a), (2.39b), (2.46) and (2.47) hold. ■

Chapter 3

Computer Simulations of
Self-Tuning Control Algorithms

3.1   Introduction

S imulation is the repetitive solution of a set of mathematical equations that describe a dynamic process [6, Banman]. The rapid development of computer technology accompanied by declining costs has rendered simulation an important engineering design tool. The feasibility of this is demonstrated in [6, Banman] in the context of industrial design. Case studies show how simulations can influence design plans, result in the detection of unexpected errors and the realization of added benefits, while reducing the overall cost of large scale projects. Successful simulations rely on realistic models followed by more practical considerations such as sampling time, choice of computer systems, etc. In this study however, sampling times are not an issue since the mathematical models used are discrete-time models. The choice of computer is also not an issue. Simulations were performed on the only available computer, namely, an AMDAHL 5850 with a 5060 accelerator running under OS/MVS. The simulations presented here do not have a practical application setting. However, a host of practical applications are summarized in the recent survey of [3, Åström]. There, it is mentioned that several hundred simulations in self-tuning control have been performed to date. However, most of them are dedicated to particular applications, and it is felt that simulations detached from applications as such are also valuable, both theoretically and practically.

A direct comparison of the performances of the much studied SG, QLS, and MLS algorithms in self-tuning control are presented in this chapter. Such a comparison, even from the point of view of a simple inspection of graphs, is hard to come by in the literature and may be non-existent. The simulations presented here only illustrate various aspects of the theory presented in Chapter 2, and do not represent an exhaustive investigation of the many interrelationships of the variables involved.

3.2 Pseudo-random disturbance sequences

Pseudo-random number generators are an essential ingredient of the simulations that were performed. The generators used form part of the International Mathematical and Statistical Library (IMSL) of Fortran subroutines. In particular, the subroutines GGUBS and GGNML were used to generate uniformly and normally distributed deviates respectively. The randomness and distribution of these deviates play a key role in the results themselves and in their interpretation. Tests were therefore performed to assess their quality.

Figures 3.1 (a), (c), and (e) show the un-normalized empirical densities of 500, 3000, and 9000 $N(0,1)$ deviates. The horizontal scale represents 100 equal subdivisions of the closed interval in which the deviates lie. Figures 3.1 (b), (d), and (f) are the corresponding plots of the sample path autocorrelation functions while figures 3.2 (a) - (f) are identical plots to those in 3.1 (a) - (f) but for $U(-\frac{1}{2},\frac{1}{2})$ deviates. Both of the figures show the expected improvement in the distribution and lack of correlation as the number of deviates increases and indicate the quality of approximation of their theoretical counterparts.

Besides the visual display of the densities in Figures 3.1 and 3.2, the Kolmogorov one-sample test (also known as the Kolmogorov-Smirnov test) was used to compare the theoretical distributions $N(0,1)$ and $U(-\frac{1}{2},\frac{1}{2})$ to some empirical ones. For each test the null hypothesis H_0 that the deviates come from $N(0,1)$ and $U(-\frac{1}{2},\frac{1}{2})$ populations was tested against the two-sided alternative A that they do not. That is, if $F_n(\cdot)$ denotes the empirical distribution of a sample of size n and $F_0(\cdot)$ is a given continuous distribution, the null hypothesis

$$H_0 : F_n(x) = F_0(x)$$

was tested against the alternative

$$A : \exists x \quad \text{s.t.} \quad F_n(x) \neq F_0(x).$$

The statistic used is $\sqrt{n}D_n$ where

$$D_n = \sup_x |F_n(x) - F_0(x)|.$$

For each sample size n, various tests have been conducted and the value of the statistic $\sqrt{n}D_n$ with associated probabilities α are shown below. The table applies to both the $N(0,1)$ and $U(-\frac{1}{2}, \frac{1}{2})$ deviates.

n	$\sqrt{n}D_n$	α		n	$\sqrt{n}D_n$	α
500	0.84697	0.46994		3000	0.62956	0.82289
	0.61764	0.84010			0.52920	0.94215
	1.1736	0.12722			0.83370	0.49041
	0.49329	0.96807			0.72812	0.66406
	0.95508	0.32129			0.60206	0.86154
	0.60167	0.86207			0.49329	0.96807
	0.43334	0.99189			0.81121	0.52601
	0.88478	0.41412			0.55179	0.92101
	0.57316	0.89771			1.1091	0.17074
	0.60910	0.85201			0.62938	0.82315
1000	0.86976	0.43581		9000	0.45699	0.98508
	1.0071	0.26251			0.44322	0.98941
	0.40111	0.99708			0.51373	0.95448
	1.3255	0.059564			0.65056	0.79114
	0.87684	0.42549			0.44870	0.98781
	0.93389	0.34767			0.65926	0.77754
	0.60701	0.85487			0.50173	0.96283
	0.36402	0.99938			1.0800	0.19386
	0.74302	0.63893			1.3633	0.04859
	0.78163	0.57430			0.80003	0.54410

Table 3.1 Statistical Results on Disturbance Sequences

For each test, the conclusion is that the test is significant $p = \alpha$, i.e. the probability that H_0 is false is $1 - \alpha$. The results above show that one should not be overly confident that the deviates are distributed as desired.

3.3 Algorithm implementation

The results of Chapter 2 involve three algorithms: the SG, the QLS, and the MLS algorithms. The objective of the experiments, i.e. the simulations, was to produce accurate and reliable information in a reasonable amount of time. Consequently, an effort was made to code numerically stable implementations of the algorithms but efficiency was not optimized. The programs were written in standard (level 77) Fortran and were carefully documented. In the author's view they constitute a first attempt in creating a simulation package of which the written programs form the innermost shell.

The simplicity of the SG and the QLS algorithms is what makes them appealing and there is little to be said about their implementation. On the other hand, the matrix gain MLS algorithm is numerically problematic. Experience in coding equation (2.30b) indicates that it is sensitive to computer roundoff and suceptible to accuracy degradation due to the differencing of two positive terms [12, Bierman]. This degradation can lead to results which cease to be meaningful, such as loss of positivity or symmetry of the error covariance matrix. It is well established that square root filtering, the reformulation of the filtering problem in terms of a square root of the error covariance matrix, improves numerical accuracy and leads to numerically stable algorithms. Such an algorithm, credited to Potter, was used to update a square root S of the matrix $\mathbf{R}$ in (2.30b), or more generally, of $\mathbf{R}$ in (4.18). With $\mathbf{R} = SS^T$, one can verify (see [12, Bierman]) that the recursive equation for S corresponding to (4.18) for $\mathbf{R}$ is

$$S(t-1) = \frac{1}{\sqrt{\lambda(t-1)}} \left[S(t-2) - \frac{\sqrt{D(t-1)}}{1 + \sqrt{D(t-1)}} \mathbf{q}(t-1)\mathbf{v}^T(t-1) \right]$$

where

$$\mathbf{v}(t-1) \triangleq (\varphi^T(t-1)S(t-2))^T,$$
$$D(t-1) \triangleq \lambda(t-1) + \mathbf{v}^T(t-1)\mathbf{v}(t-1),$$
$$\mathbf{q}(t-1) \triangleq \frac{S(t-2)\mathbf{v}(t-1)}{D(t-1)},$$

and $\lambda(t)$ a possibly time varying forgetting factor (see section 4.3). The error covariance $\mathbf{R}$ may be obtained at any instant by forming the product SS^T. In the rest of this chapter, reference to the MLS algorithm shall imply use of Modification 1.

3.4   Simulation Descriptions

The simulations presented here illustrate the ability of the self-tuning controller to stabilize an unstable system and simultaneously enable it to track a given reference sequence. In most of the plots that follow, 2 graphs are overlayed on a single set of axes. The solid line graphs pertain to the closed loop self-tuning system while those with a dashed line to the performance of the system under the assumption of known parameters (i.e. no adaptation). The examples shown vary only in certain variables in order to isolate their effect. Sample paths of 1000 samples each were simulated and for each example, the self-tuning output $y(t)$, self-tuning input $u(t)$, self-tuning average squared tracking error $ASTE = 1/N \sum_{i=1}^{N} (y(t) - y^*(t))^2$, parameter estimates $\hat{\vartheta}_t(t)$, and output sample path autocorrelation function $\hat{R}_y(\tau)$ were plotted.

A case under consideration is that of $y^*(t) \equiv 0$. Via (2.6) and (2.11), $y(t) = w(t)$ since for the unit delay case which was simulated, $F(z) = 1$. Then, theoretically, in closed loop, the non-self-tuning controlled output sequence is uncorrelated since $\{w(t)\}$ is while the output of the self-tuning algorithm is at least asymptotically uncorrelated. The autocorrelation function displays the degree of correlation of the sequence plotted. Further, in the case of known parameters and under minimum variance control, the output $y(t)$ is a linear combination of $x_0, w(1) \ldots w(t)$, the determining uncorrelated random variables for $\mathcal{F}_t$. If these are normally distributed then so is $y(t)$. This can be verified by obtaining values of $y(t)$ for a fixed t corresponding to different sample paths. This leads to the question of what is the distribution of $y(t)$, for fixed t, in the self-tuning case. Empirical distributions of cross sectional self-tuning output sequences are examined and plotted in various examples presented hereafter and the Kolmogorov one-sample test is performed.

Example 3.1

This example shows how a self-tuning controller stabilizes an unstable system and guides it to track the zero reference sequence i.e. it is an example of regulation about 0. The delay is assumed to be unity ($d = 1$) while the $A(\cdot), B(\cdot)$, and $C(\cdot)$ polynomials are

$$A(z) = 1 + 2z, \quad n = 1,$$
$$B(z) = 1 + \frac{1}{2}z, \quad m = 1,$$
$$C(z) = 1 - \frac{1}{5}z, \quad \ell = 1.$$

It is assumed that the controller knows (i.e. may explicitly depend on) the delay d and upper bounds n', m', and ℓ' for n, m, and ℓ respectively. These are $n' = m' = \ell' = 2$. Figures 3.3, 3.4, and 3.5 show simulation results for the SG, QLS, and MLS (Modification 1) algorithms with the associated stability and optimality results found in theorems 2.5, 2.6, and Corollary 2.7a respectively. The assumptions for their conclusions have all been respected.

The disturbance process $\{w(t)\}$ used is i.i.d. using a $N(0,1)$ pseudo-random number generator. The initial condition is $x_0 = 0$ while the initial parameter estimate is chosen as

$$\hat{\vartheta}(0) = [-5 \ \ 5,6 \ \ 1-6, 11-1]^T$$

to display an initial transient behaviour. Each of the figures 3.3, 3.4, and 3.5 are composed of 11 graphs, labelled (a) - (k). In (a) the self-tuning output is displayed, (b) displays the self-tuning input, and (c) the self-tuning ASTE. In (d) - (j) are found the seven estimated parameters of the predictor model while (k) shows a plot of the output sample path autocorrelation function. The same disturbance sequence is used to simulate all three algorithms. In the SG and QLS cases the actual predictor parameters are

$$\vartheta^\circ = [-1.8 \ \ 0,1 \ \ 0.5 \ \ 0,0.2 \ \ 0]^T$$

while in the MLS case

$$\vartheta^\circ = [-2 \ \ 0,1 \ \ 0.5 \ \ 0,0.2 \ \ 0]^T.$$

These are overlayed in the plots of the estimated parameters for comparison. Further, in the MLS case, the initial error covariance matrix is chosen as $100I$ (I is the identity matrix) to ensure rapid initial convergence while the constants K_1 and K_2 were chosen at 10 and 0.9 respectively.

Stabilizability is demonstrated by all three algorithms. They all give an output which eventually varies about the tracking sequence with a variance close to 1 although the SG and QLS algorithms display a large initial transient behaviour which can be deemed unacceptable. This transient is due to the combination of the choice of $\widehat{\vartheta}(\cdot)$ and the fact that the system is unstable. For example, simulation of the unstable case with the QLS algorithm and $\widehat{\vartheta}(0) = \vartheta^\circ$ does not result in a transient. It is surprising that the performance of the SG algorithm is superior to that of the QLS algorithm while given experience reported in the literature, it is not at all unexpected that the MLS algorithm is far superior to both the SG and QLS algorithms. Its rate of convergence is faster and the transient much smaller. ∎

Example 3.2

Only the MLS algorithm is simulated. The reference sequence is

$$y^*(t) = 2\sin\left(\frac{2\pi t}{500}\right) \qquad t = 1, 2, \ldots, 1000$$

and the disturbance sequence $w(t) \sim N(0, \frac{1}{5})$. With only these two changes from example 1, Figure 3.6 shows good stabilization and tracking performance. Figure (3.6)(b) shows the non-self-tuning system output while figures 3.6(a) and 3.6(c) - (ℓ) are the same kinds of plots as the plots (a) - (k) in the figures pertaining to example 1. ∎

Example 3.3

Here the performances of SG, QLS and MLS algorithms on a stable system are examined. The only change from example 1 is in the $A(\cdot)$ polynomial which was changed to

$$A(z) = 1 + 0.5z, \qquad n = 1.$$

Figures 3.7, 3.8, and 3.9 show the performance of the QLS, MLS, and SG algorithms respectively. It is evident that the initial transient has now almost disappeared and the difference in performance among the three algorithms is not as different as in example 1. The QLS algorithm is still worse than the SG while the MLS algorithm wins marginally over the SG. Note how the output sample path autocorrelation function approaches that of the non-adaptive output (which is equal to $w(t)$ in this example). This was not the case in example 1. Further, for the SG case, 500 sample paths of 1000 samples each were simulated and the values of $y(300), y(600)$ and $y(900)$ were recorded. Their empirical un-normalized densities appear in figure 3.9 (ℓ) - (n). Comparison with the density of figure 3.1(a) shows that the output has a sample distribution resembling that of a normally distributed random variable. The Kolmogorov one-sample test was used to test for normality of $y(300), y(600)$, and $y(900)$ as well as of $w(300), w(600)$, and $w(900)$. These tests were performed using the empirical means and variances of the above quantities and the results are tabulated below.

	$\sqrt{n}\,D_n$	α	μ	σ^2
$w(300)$	0.47459	0.97792	0.05985	1.0854
$y(300)$	0.44880	0.98778	0.02238	1.2716
$w(600)$	0.55593	0.91674	-0.02829	0.90104
$y(600)$	0.74825	0.63012	-0.04799	1.2734
$w(900)$	0.56560	0.90630	0.11039	1.0694
$y(900)$	0.61255	0.84724	0.09669	1.3666

Table 3.2 Kolmogorov one-sample test results

These results inspire a relatively high degree of confidence that the random variables tested are normally distributed. As was mentioned at the beginning of this section, if x_0 is normally distributed, such a result is expected for the non-self-tuning case since then the output is just a linear combination of normally distributed random variables. In the self-tuning case, the output tends asymptotically in a sample mean square sense to such a linear combination giving reason to believe the validity of the aforementioned conclusions of normality. ∎

It is noteworthy that while in all these simulations — and in others not presented here — the parameters converged, convergence to their true values was never observed, nor was convergence to a fixed multiple of them.   Persistent excitation induced by setting $y^*(t)$ to a random variable $x \sim U(-a, a)$ for various values of a yielded, as before, negative results as far as consistent parameter estimation is concerned. Thus the conclusions of theorems 2.11 and 2.12 could not be verified. However, since the system is overmodeled, the parameters can converge to values for which there exist *common factors*, as proved in Theorem 19 of Becker et. al. The true optimal control is,

$$u = \frac{(A \cdot B)}{B} y = \frac{1.8z}{5.5 - 0.5 - 3.8z^2} y$$

and the simulations seem to converge to,

$$u = \frac{6.8z - 5.8z^2}{5.5 - 0.5z - 3.8z^2} y$$

A rough calculation shows an approximate common factor of $(-3.2z + 4.6)$ in the numerator and denominator, justifying the theory as well as validating the simulation.

Example 3.4

This last example shows how a self-tuning controller using the SG algorithm stabilizes and regulates an unstable, under-modelled plant. Compared with example 3.1, the only change occurs in the $A(\cdot)$ polynomial which is now given by

$$A(z) = 1 + 2.1z + 1.1z^2, \quad n = 2,$$

with $n' = 1, m' = 1$, and $\ell' = 1$. The initial parameter estimate is chosen as

$$\hat{\vartheta}(0) = [-5, \; 56, \; 1]^T$$

and figure 3.10 displays the simulation results. Figure 3.10(a) shows how, after an initial large transient, the controlled output is regulated about the zero reference point. The ASTE is shown in figure 3.10(b) and the four estimated parameters in figures 3.10(c)-(f). The self-tuning output autocorrelation function is displayed in figure 3.10(g). ∎

3.5 Simulation Results

The following pages display the plots corresponding to the examples described above.

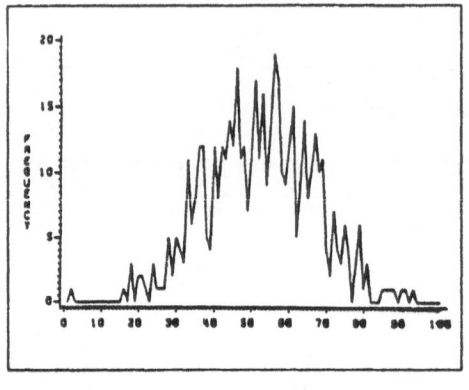

(a) Empirical density (500 samples)

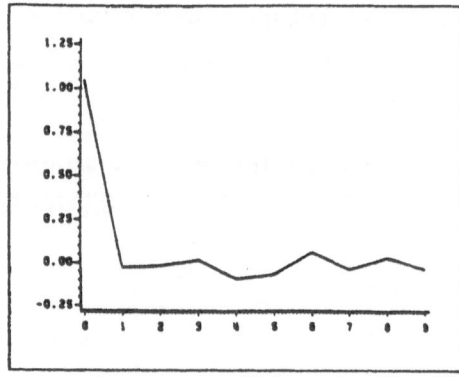

(b) Sample path autocorrelation function

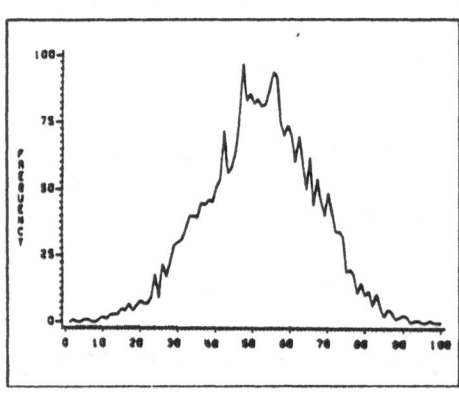

(c) Empirical density (3000 samples)

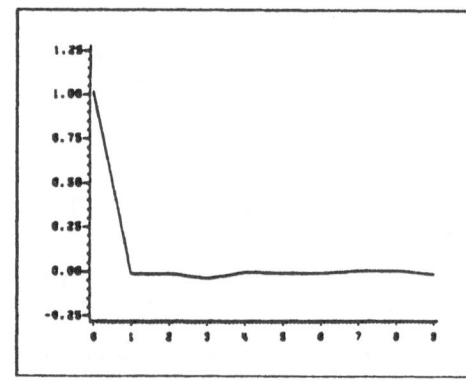

(d) Sample path autocorrelation function

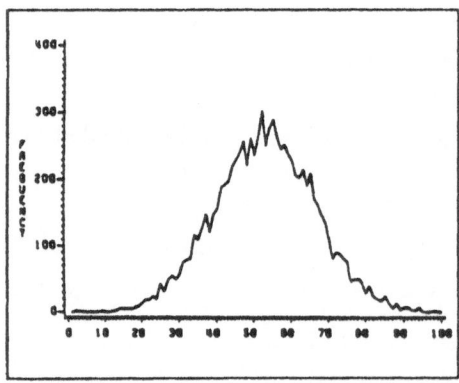

(e) Empirical density (9000 samples)

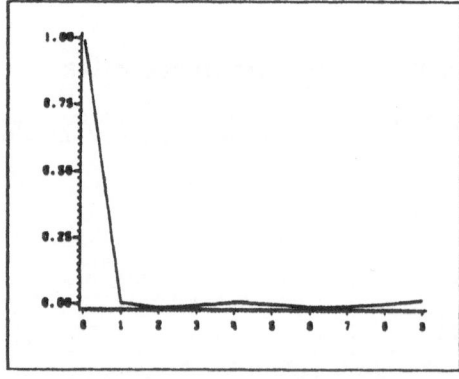

(f) Sample path autocorrelation function

Figure 3.1 Tests for $N(0,1)$ Disturbance Sequence

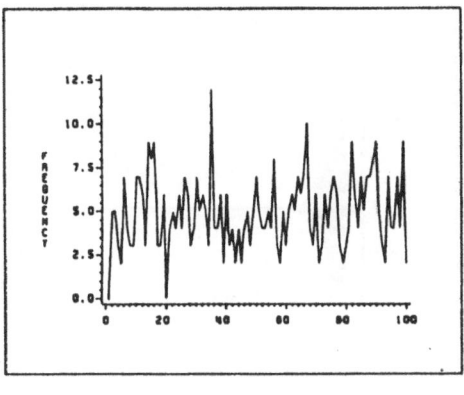

(a) Empirical density (500 samples)

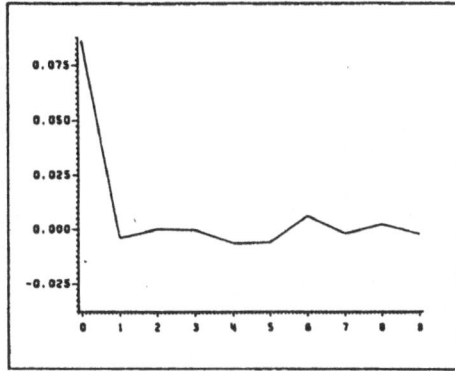

(b) Sample path autocorrelation function

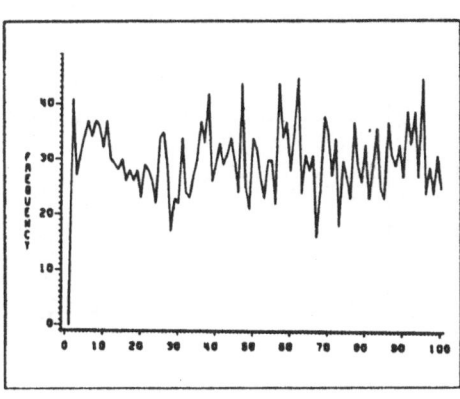

(c) Empirical density (3000 samples)

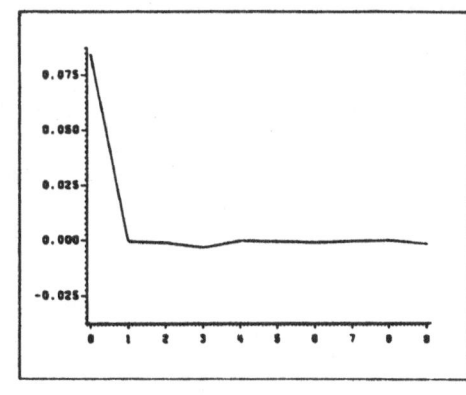

(d) Sample path autocorrelation function

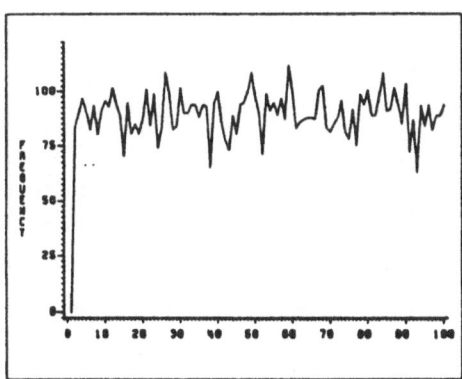

(e) Empirical density (9000 samples)

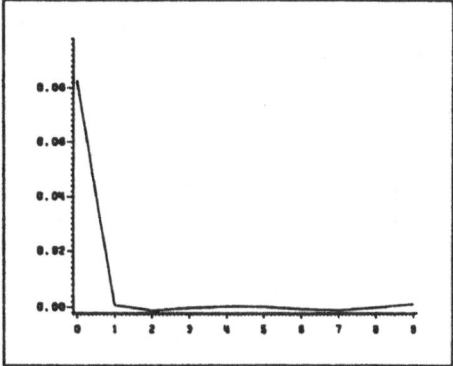

(f) Sample path autocorrelation function

Figure 3.2 Tests for $U(-\frac{1}{2}, \frac{1}{2})$ Disturbance Sequence

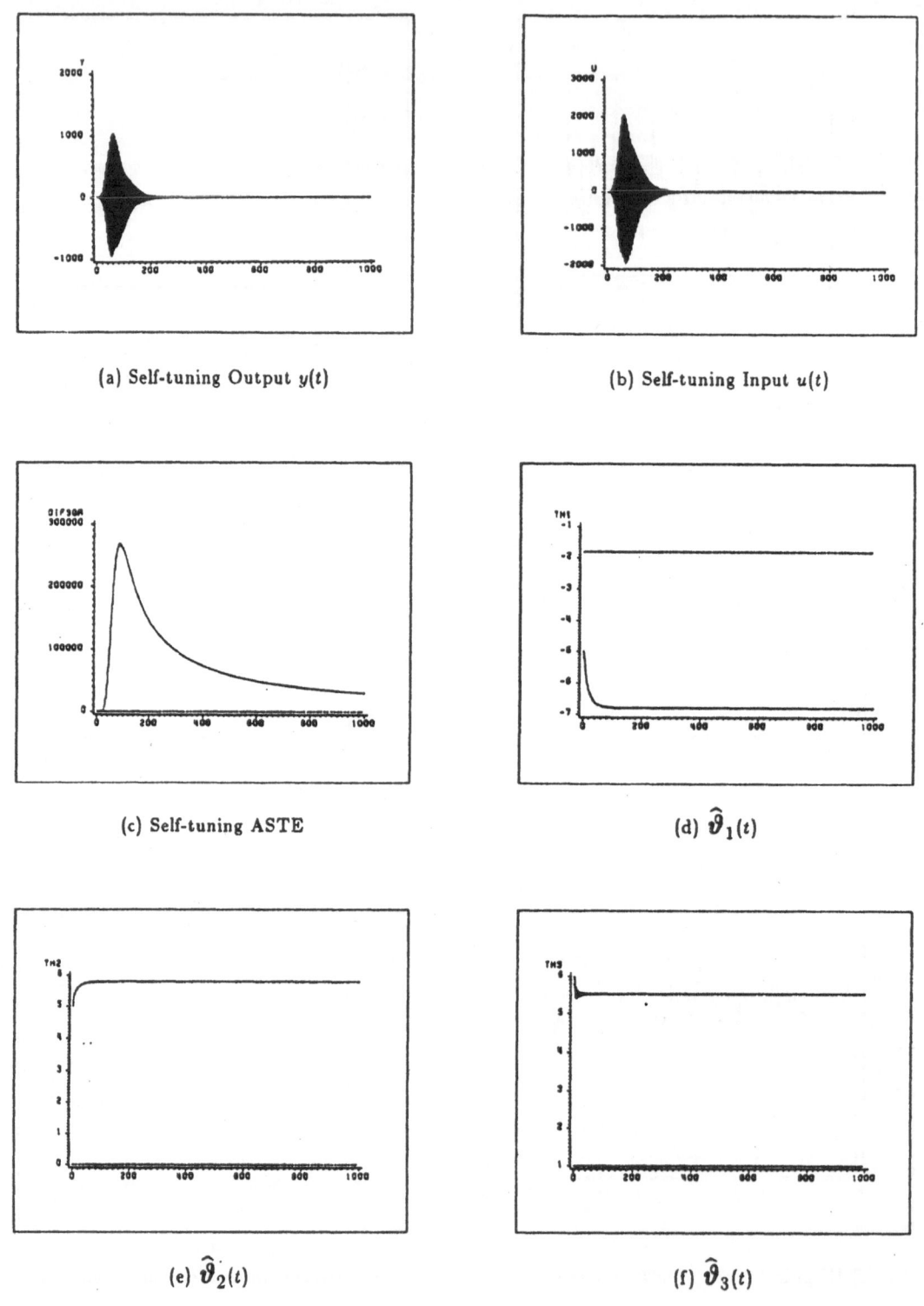

(a) Self-tuning Output $y(t)$

(b) Self-tuning Input $u(t)$

(c) Self-tuning ASTE

(d) $\widehat{\vartheta}_1(t)$

(e) $\widehat{\vartheta}_2(t)$

(f) $\widehat{\vartheta}_3(t)$

Figure 3.3   Example 3.1 (SG)

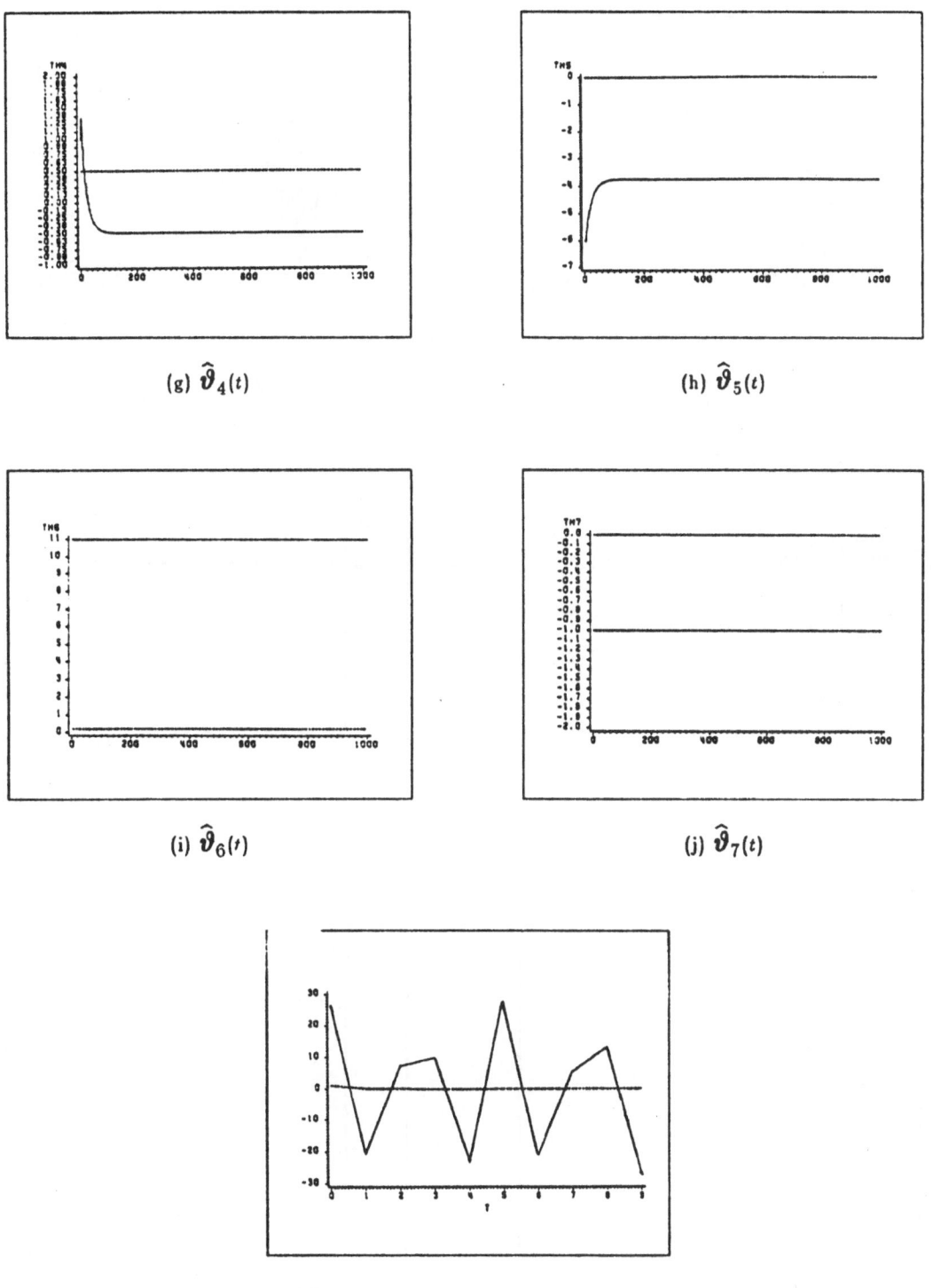

(g) $\widehat{\vartheta}_4(t)$

(h) $\widehat{\vartheta}_5(t)$

(i) $\widehat{\vartheta}_6(t)$

(j) $\widehat{\vartheta}_7(t)$

(k) Self-tuning Output Autocorrelation

Figure 3.3 Example 3.1 (SG)

44

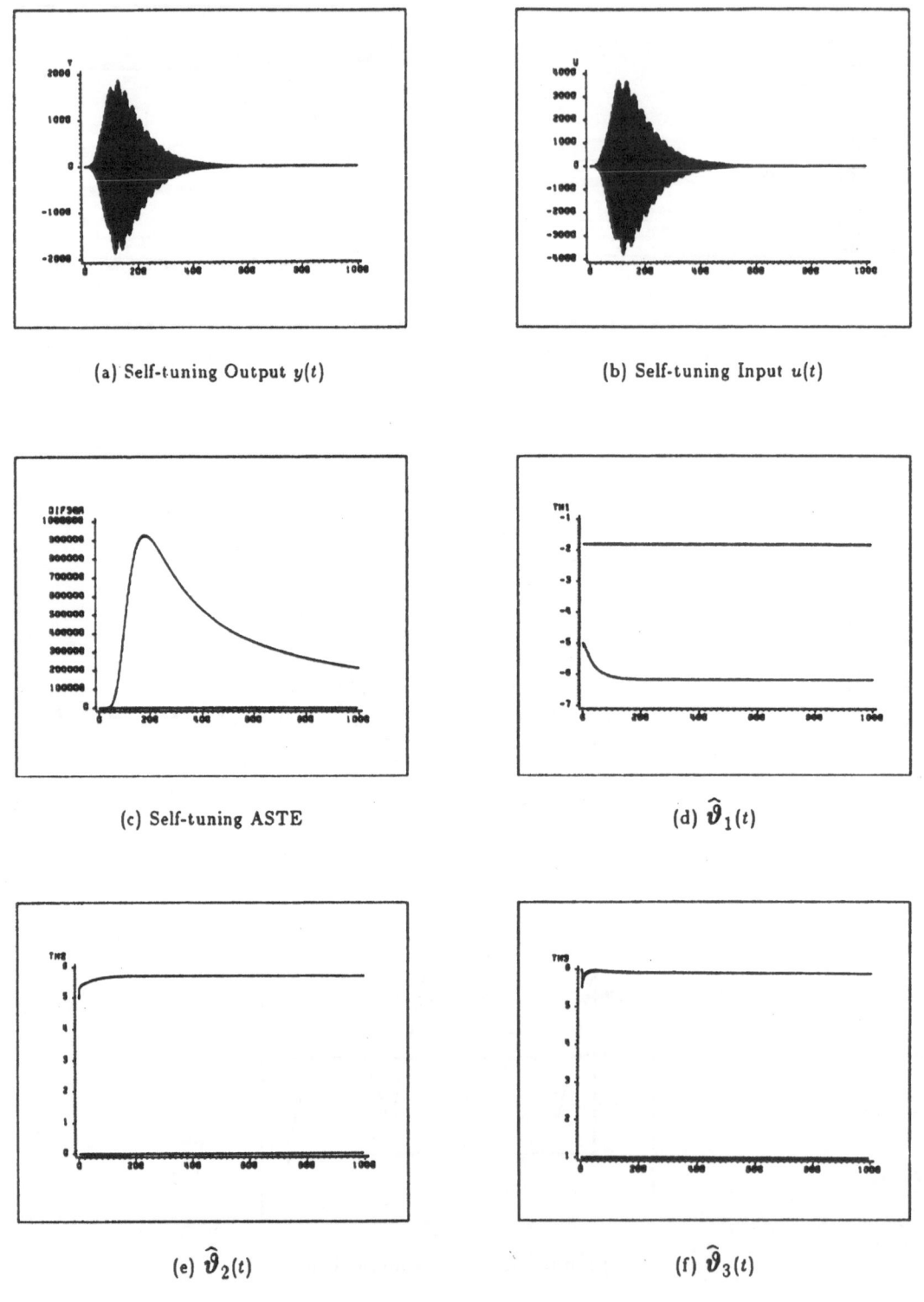

(a) Self-tuning Output $y(t)$

(b) Self-tuning Input $u(t)$

(c) Self-tuning ASTE

(d) $\widehat{\vartheta}_1(t)$

(e) $\widehat{\vartheta}_2(t)$

(f) $\widehat{\vartheta}_3(t)$

Figure 3.4 Example 3.1 (QLS)

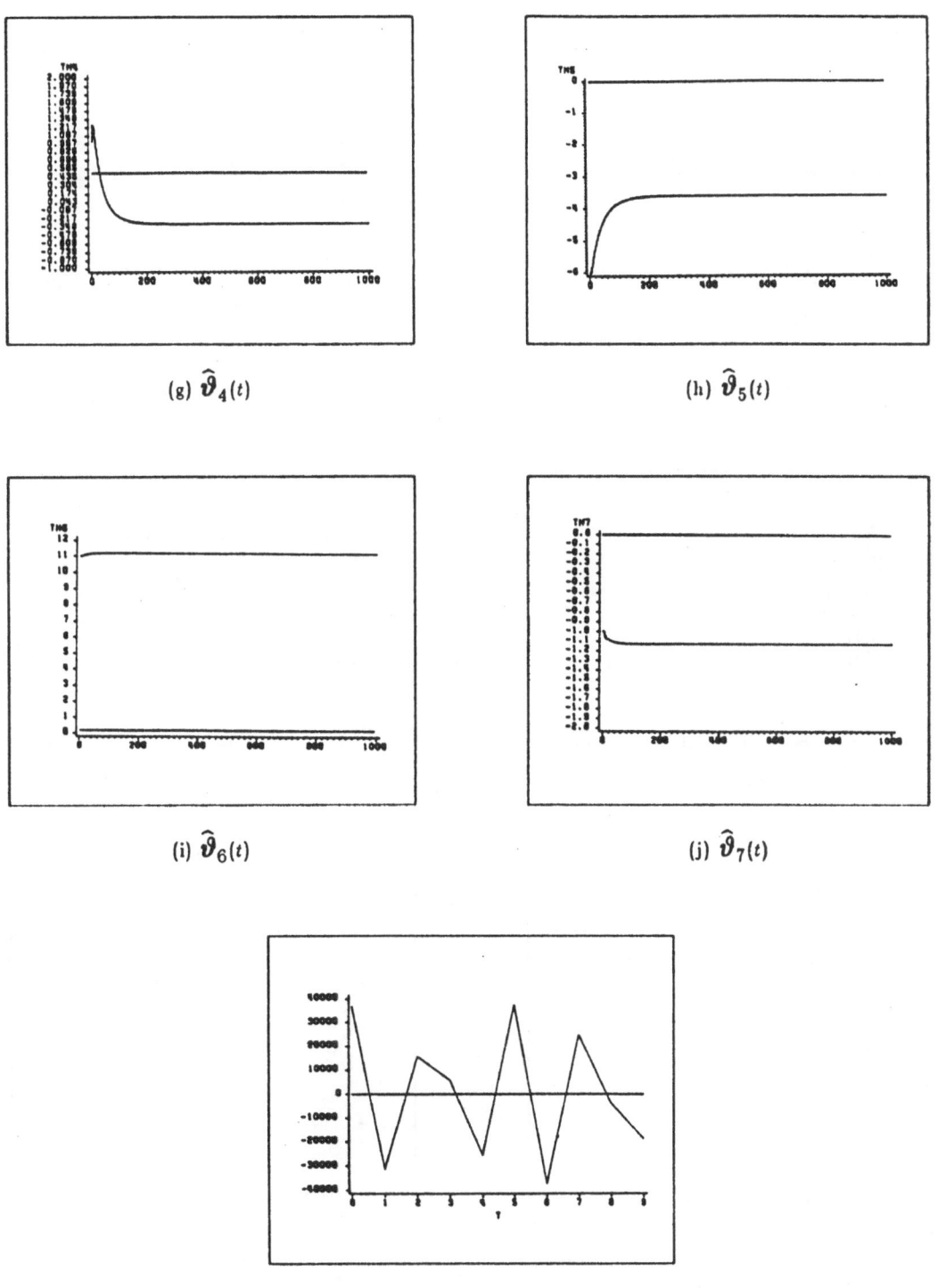

(g) $\widehat{\vartheta}_4(t)$

(h) $\widehat{\vartheta}_5(t)$

(i) $\widehat{\vartheta}_6(t)$

(j) $\widehat{\vartheta}_7(t)$

(k) Self-tuning Output Autocorrelation

Figure 3.4 Example 3.1 (QLS)

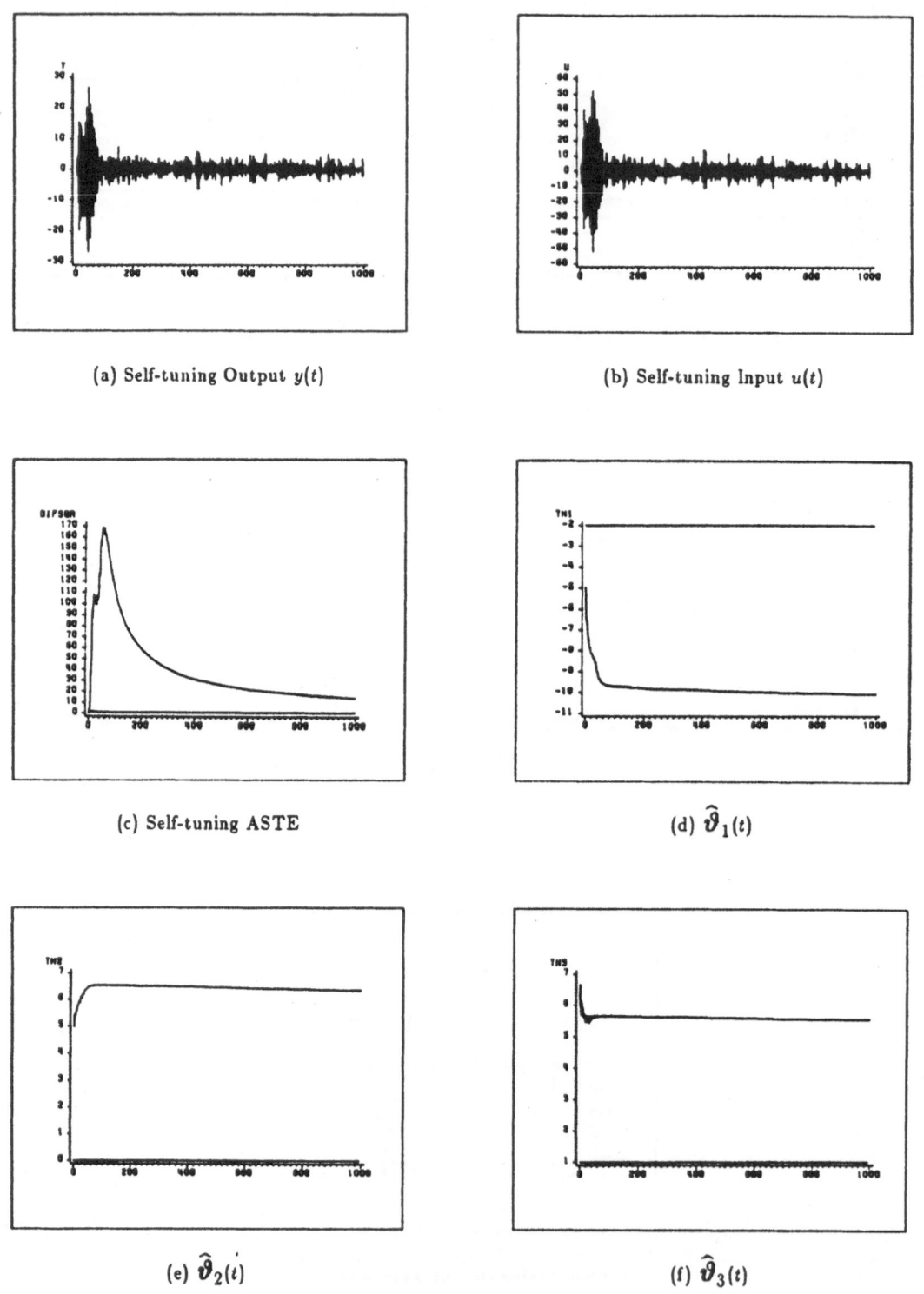

(a) Self-tuning Output $y(t)$

(b) Self-tuning Input $u(t)$

(c) Self-tuning ASTE

(d) $\widehat{\vartheta}_1(t)$

(e) $\widehat{\vartheta}_2(t)$

(f) $\widehat{\vartheta}_3(t)$

Figure 3.5 Example 3.1 (MLS)

47

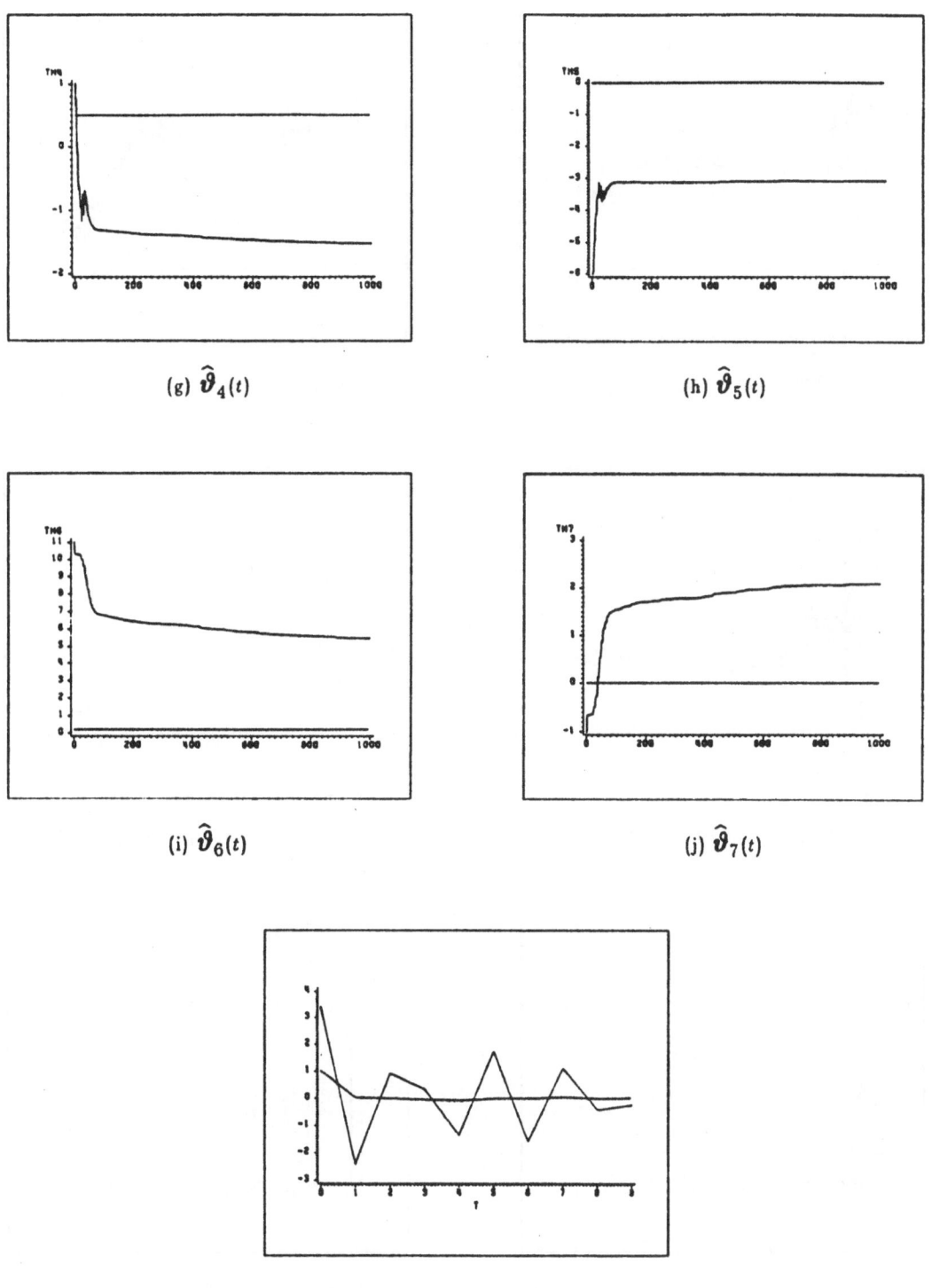

(g) $\widehat{\vartheta}_4(t)$

(h) $\widehat{\vartheta}_5(t)$

(i) $\widehat{\vartheta}_6(t)$

(j) $\widehat{\vartheta}_7(t)$

(k) Self-tuning Output Autocorrelation

Figure 3.5   Example 3.1 (MLS)

48

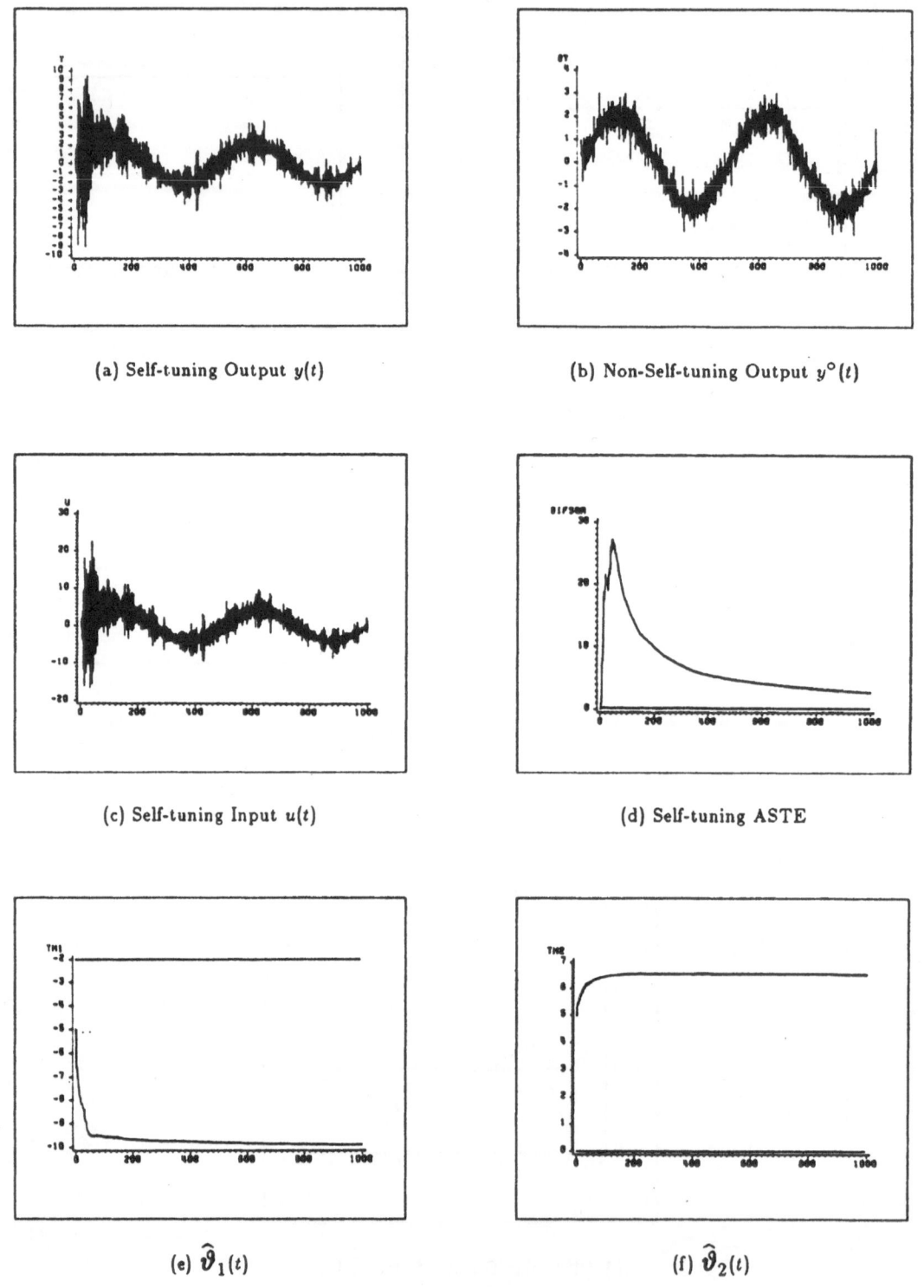

(a) Self-tuning Output $y(t)$

(b) Non-Self-tuning Output $y^\circ(t)$

(c) Self-tuning Input $u(t)$

(d) Self-tuning ASTE

(e) $\widehat{\vartheta}_1(t)$

(f) $\widehat{\vartheta}_2(t)$

Figure 3.6　Example 3.2 (MLS)

49

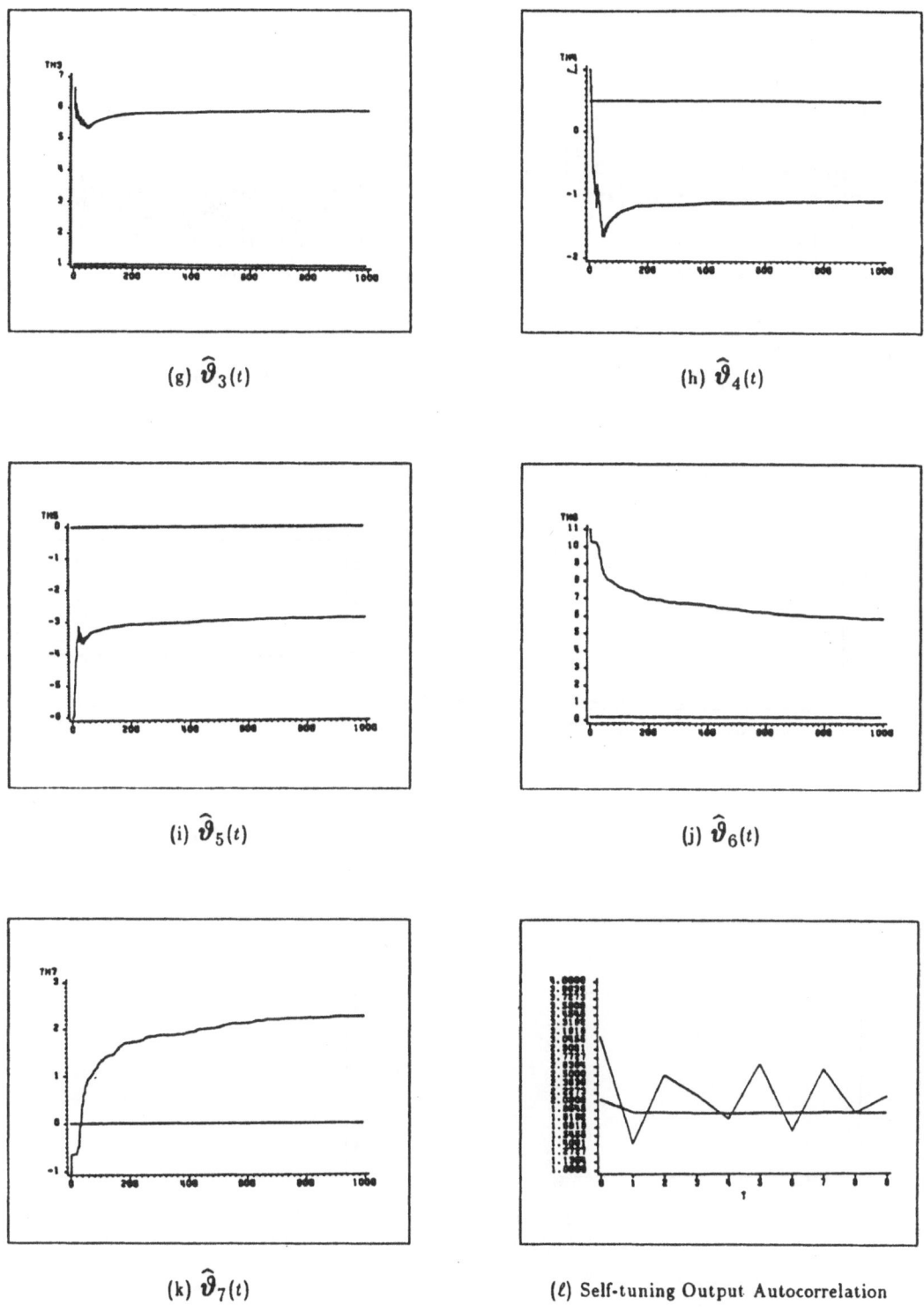

(g) $\widehat{\vartheta}_3(t)$

(h) $\widehat{\vartheta}_4(t)$

(i) $\widehat{\vartheta}_5(t)$

(j) $\widehat{\vartheta}_6(t)$

(k) $\widehat{\vartheta}_7(t)$

(ℓ) Self-tuning Output Autocorrelation

Figure 3.6 Example 3.2 (MLS)

50

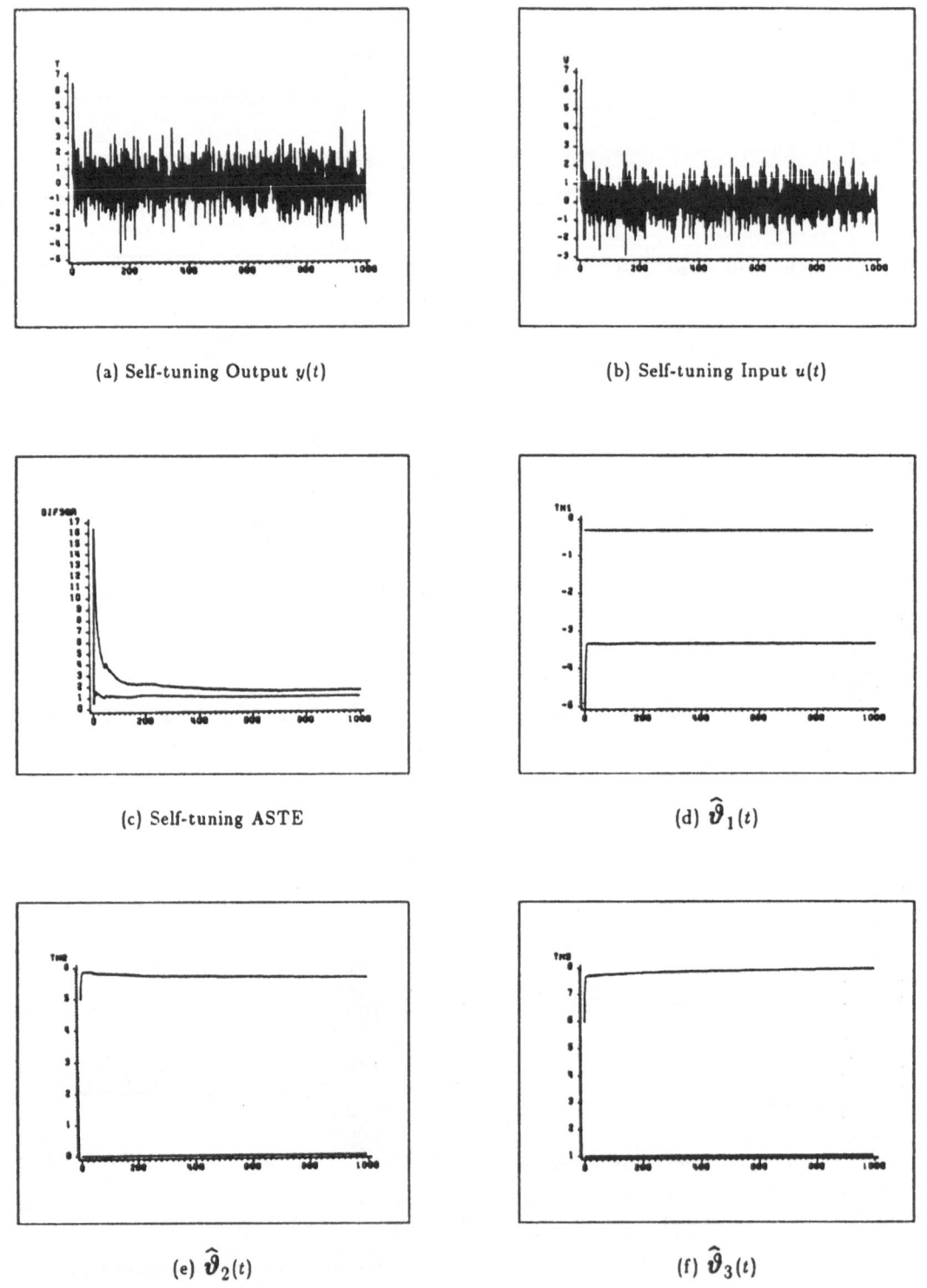

(a) Self-tuning Output $y(t)$

(b) Self-tuning Input $u(t)$

(c) Self-tuning ASTE

(d) $\widehat{\vartheta}_1(t)$

(e) $\widehat{\vartheta}_2(t)$

(f) $\widehat{\vartheta}_3(t)$

Figure 3.7 Example 3.3 (QLS)

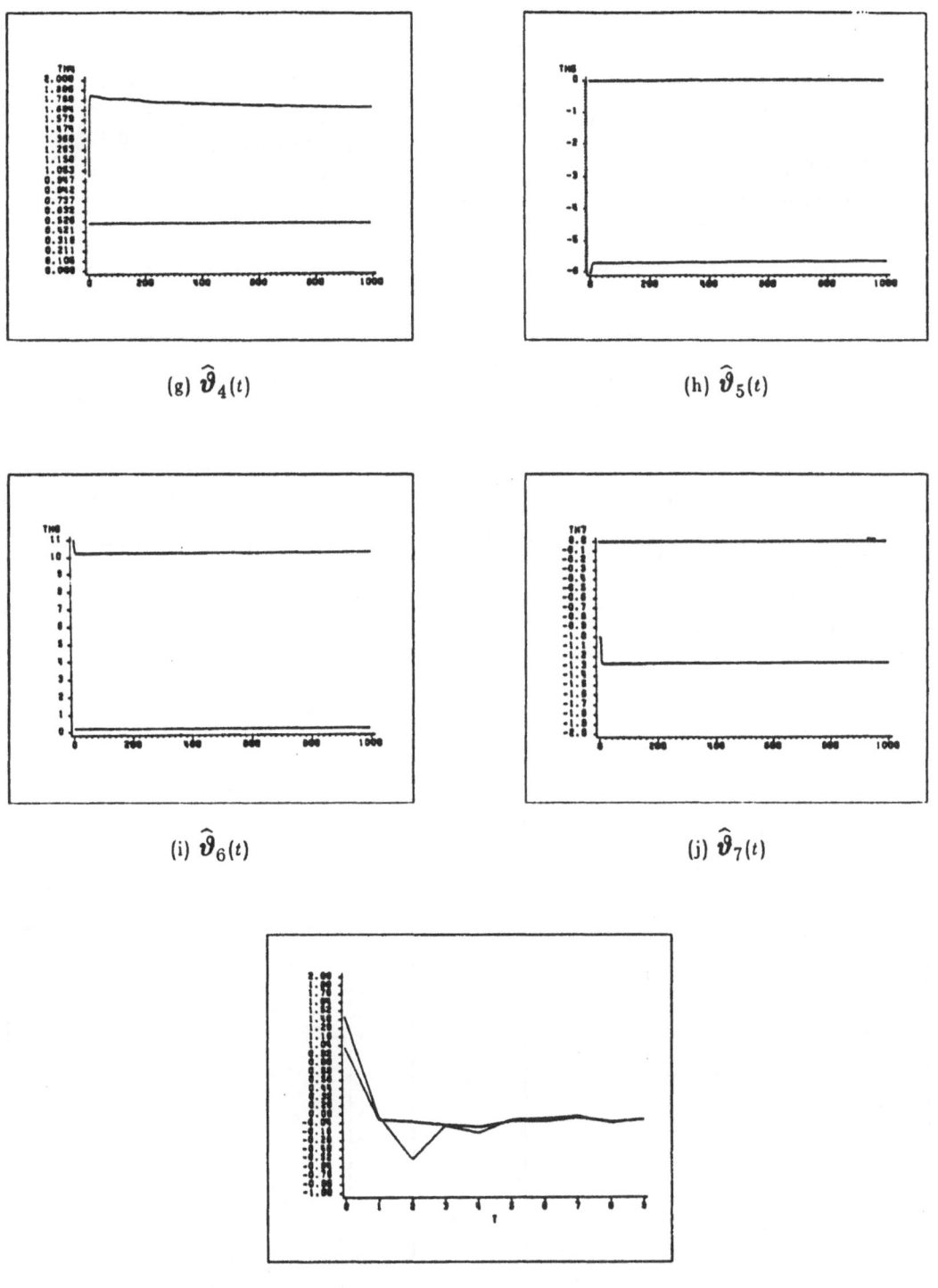

(g) $\widehat{\vartheta}_4(t)$ (h) $\widehat{\vartheta}_5(t)$

(i) $\widehat{\vartheta}_6(t)$ (j) $\widehat{\vartheta}_7(t)$

(k) Self-tuning Output Autocorrelation

Figure 3.7   Example 3.3 (QLS)

52

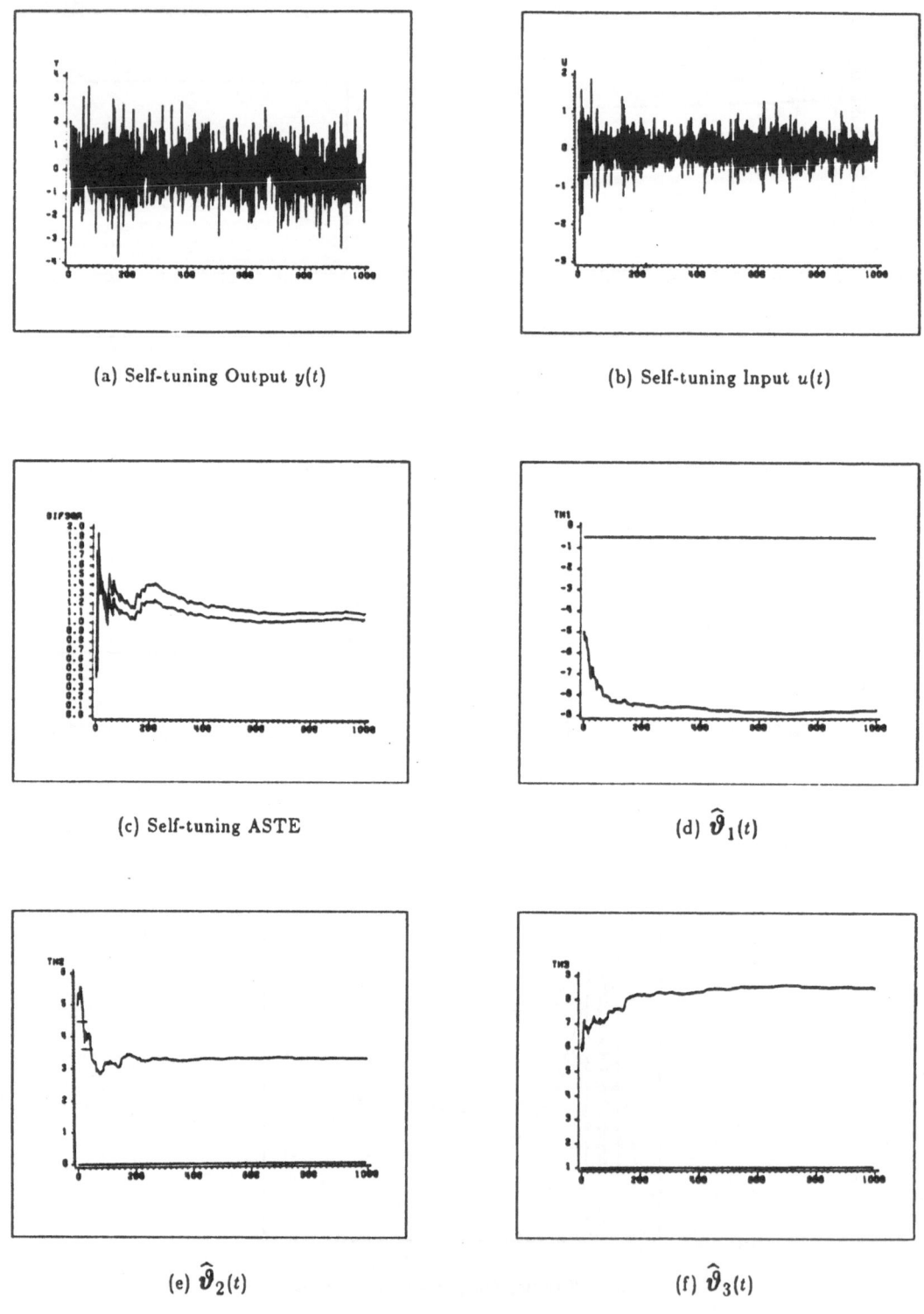

(a) Self-tuning Output $y(t)$

(b) Self-tuning Input $u(t)$

(c) Self-tuning ASTE

(d) $\widehat{\vartheta}_1(t)$

(e) $\widehat{\vartheta}_2(t)$

(f) $\widehat{\vartheta}_3(t)$

Figure 3.8 Example 3.3 (MLS)

53

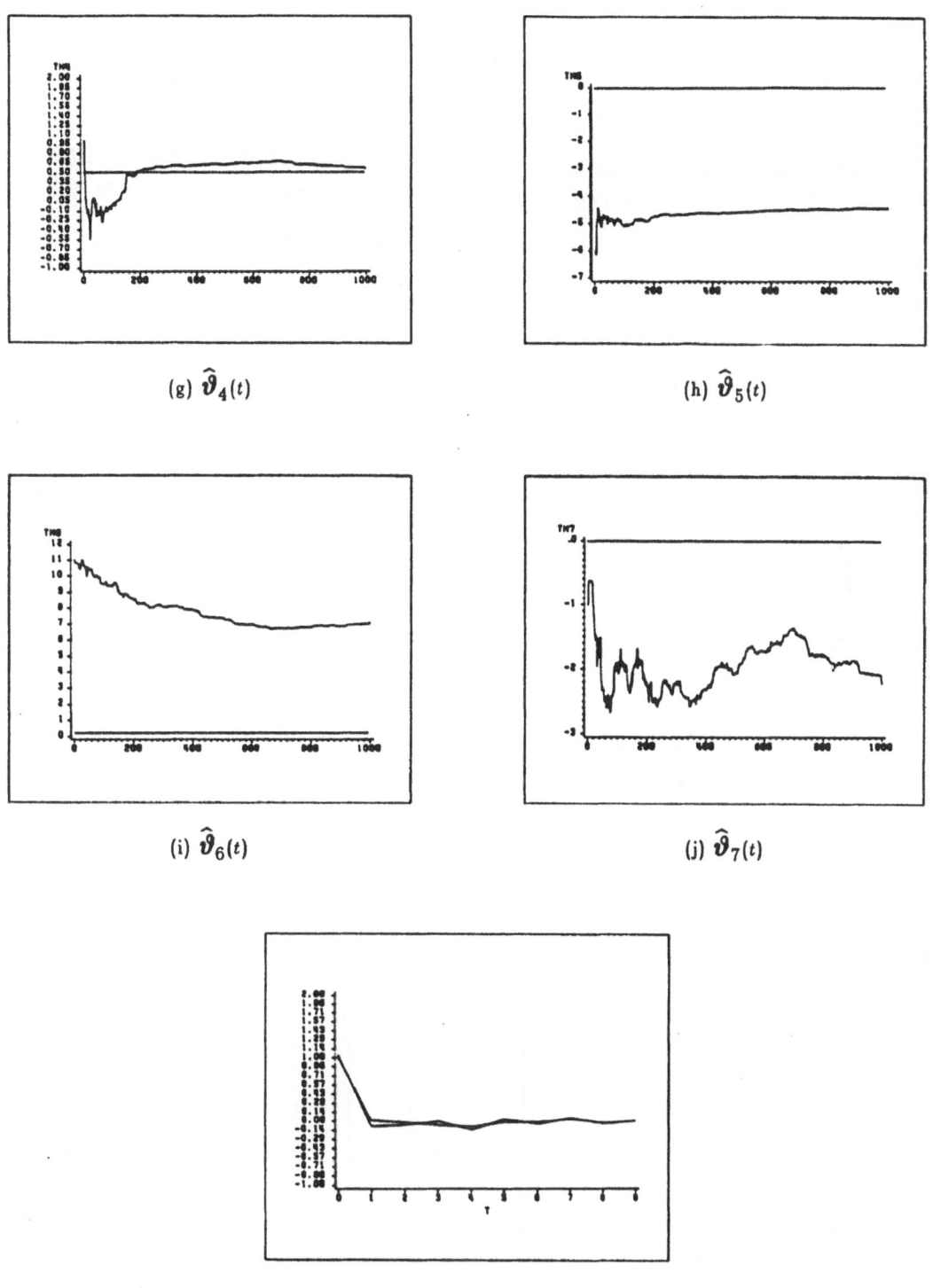

(g) $\widehat{\vartheta}_4(t)$

(h) $\widehat{\vartheta}_5(t)$

(i) $\widehat{\vartheta}_6(t)$

(j) $\widehat{\vartheta}_7(t)$

(k) Self-tuning Output Autocorrelation

Figure 3.8   Example 3.3 (MLS)

54

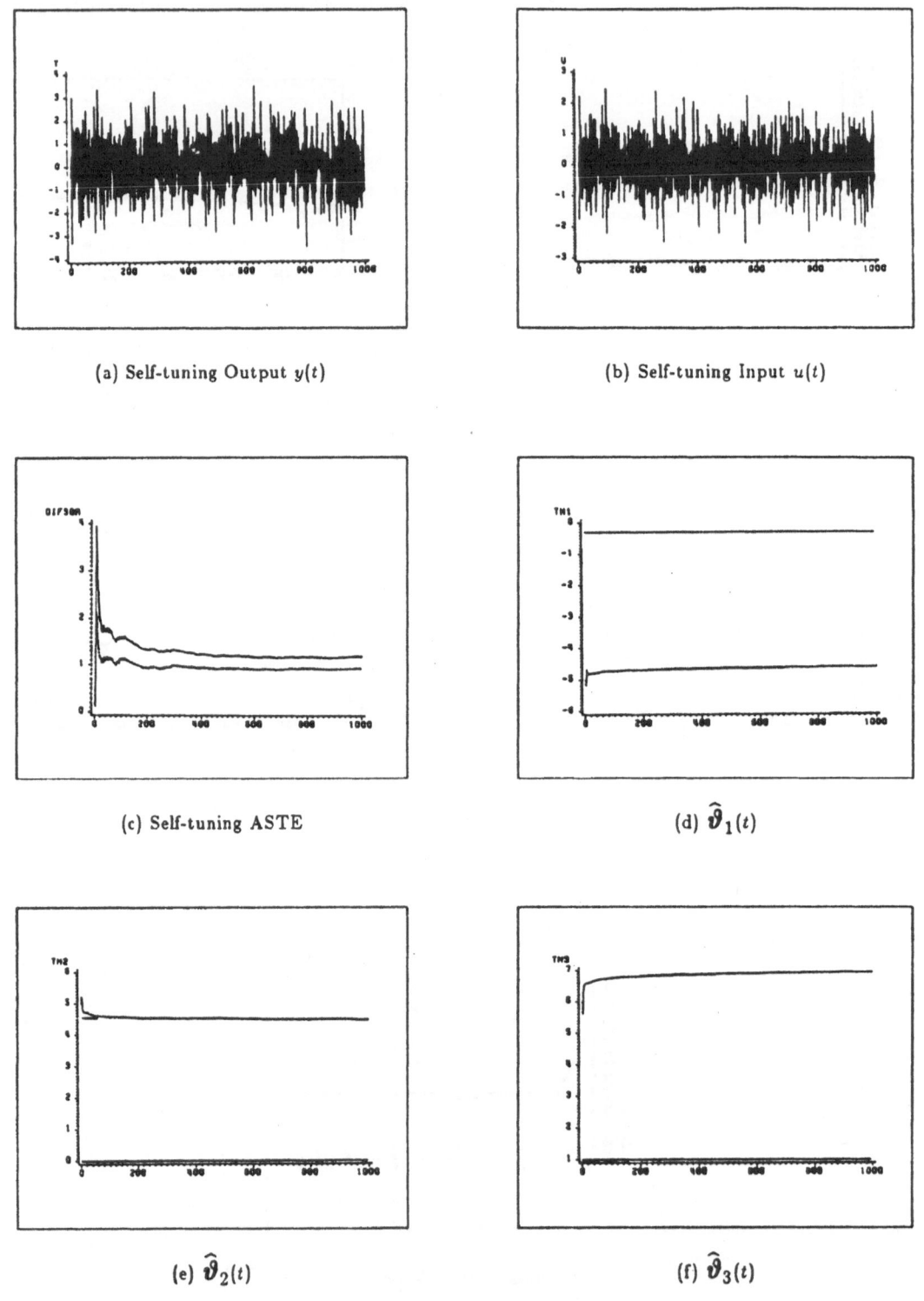

(a) Self-tuning Output $y(t)$

(b) Self-tuning Input $u(t)$

(c) Self-tuning ASTE

(d) $\widehat{\vartheta}_1(t)$

(e) $\widehat{\vartheta}_2(t)$

(f) $\widehat{\vartheta}_3(t)$

Figure 3.9   Example 3.3 (SG)

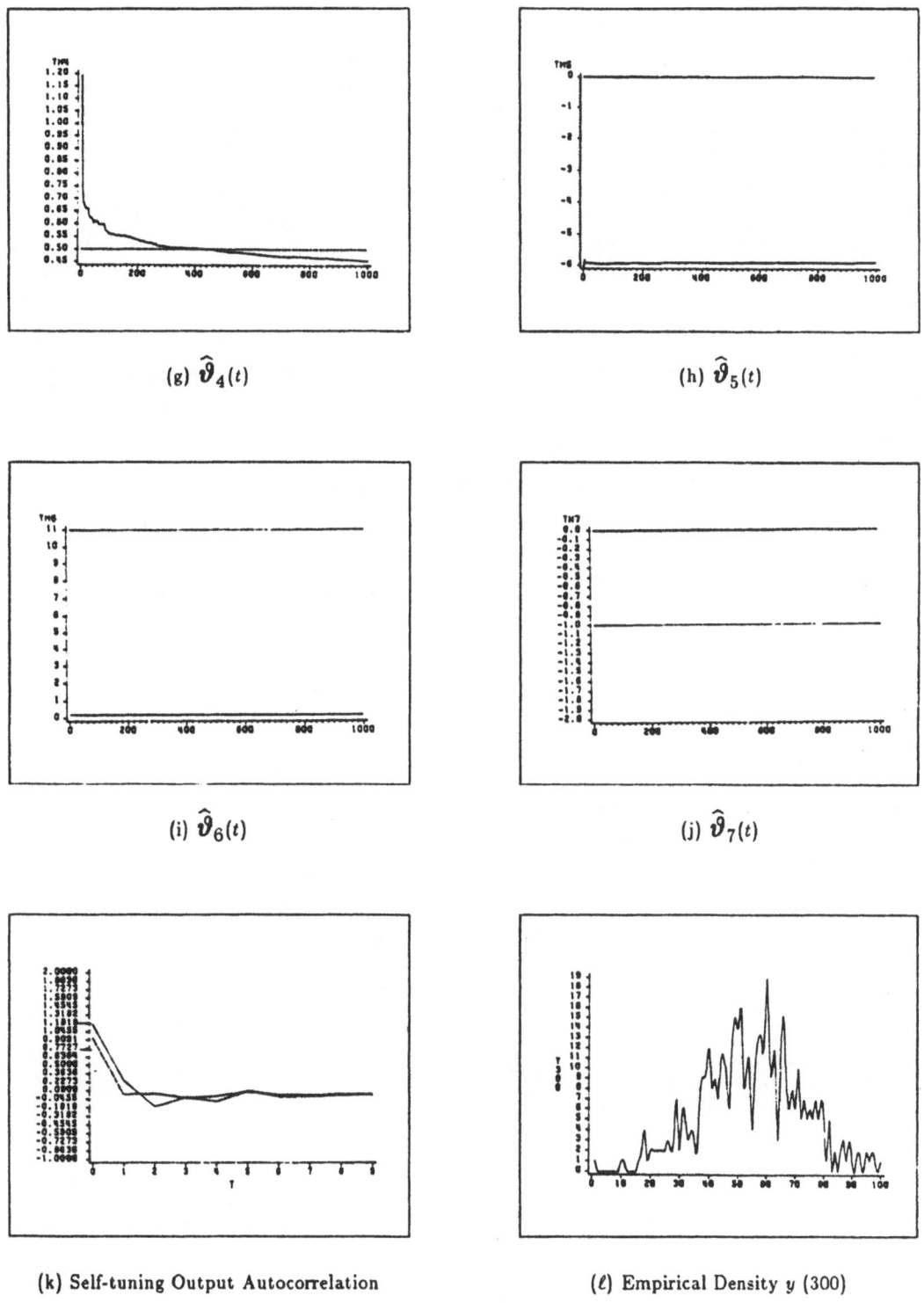

(g) $\widehat{\vartheta}_4(t)$

(h) $\widehat{\vartheta}_5(t)$

(i) $\widehat{\vartheta}_6(t)$

(j) $\widehat{\vartheta}_7(t)$

(k) Self-tuning Output Autocorrelation

(ℓ) Empirical Density y (300)

Figure 3.9   Example 3.3 (SG)

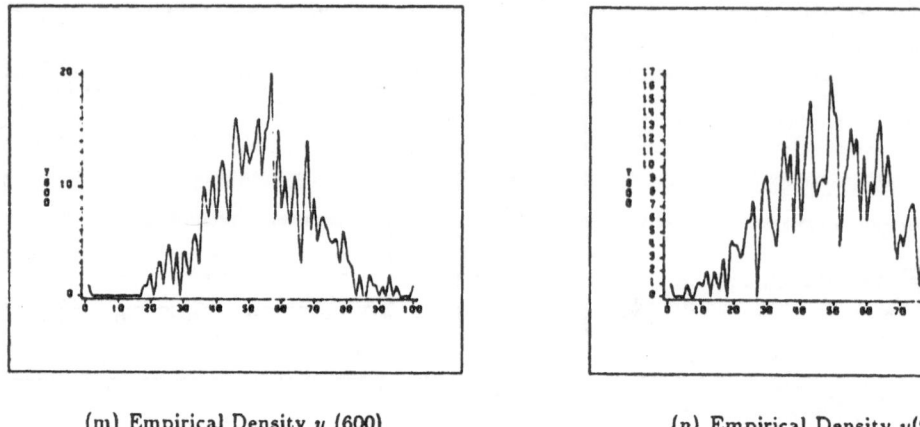

(m) Empirical Density y (600)   (n) Empirical Density $y(900)$

Figure 3.9 Example 3.3 (SG)

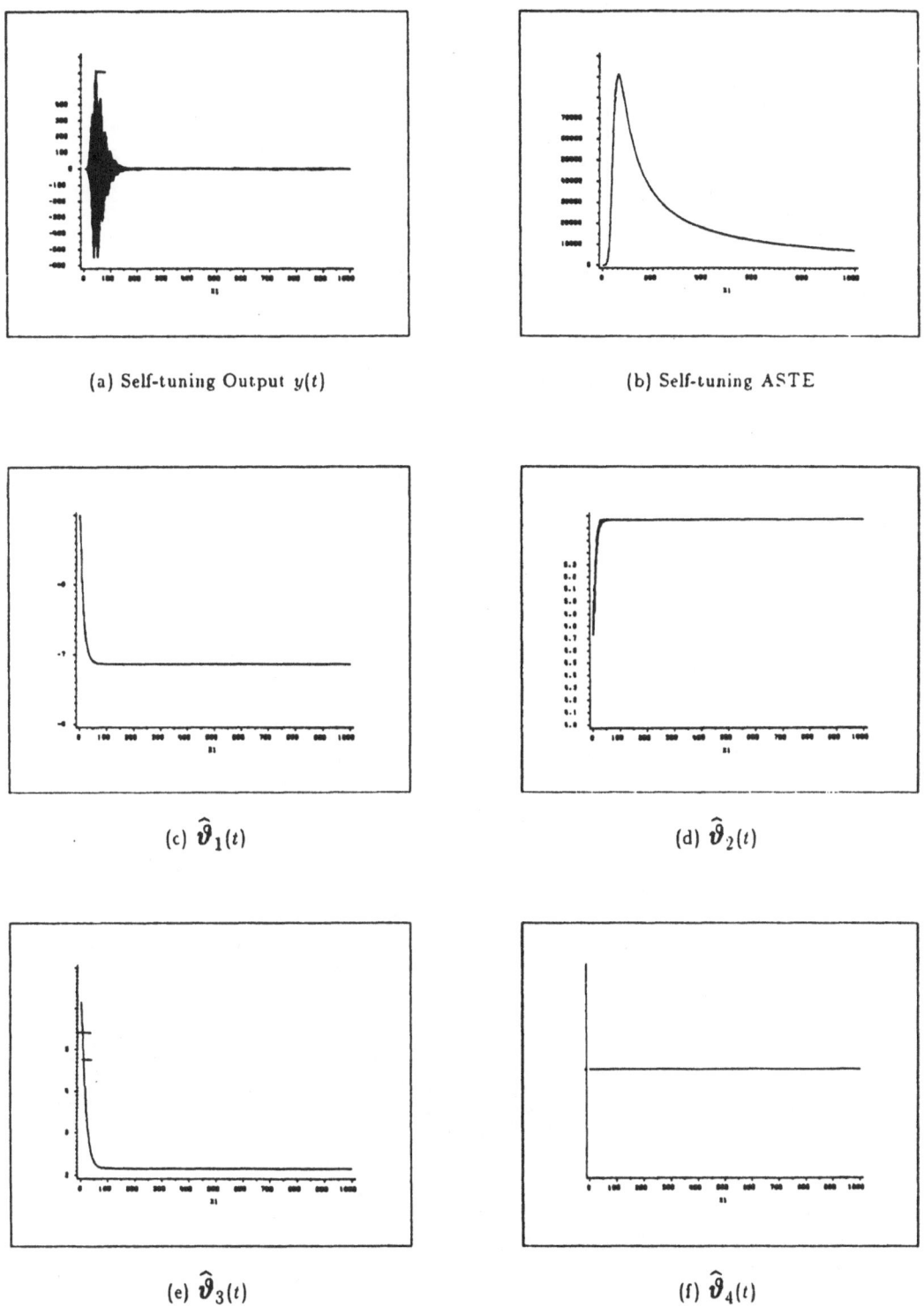

(a) Self-tuning Output $y(t)$

(b) Self-tuning ASTE

(c) $\widehat{\vartheta}_1(t)$

(d) $\widehat{\vartheta}_2(t)$

(e) $\widehat{\vartheta}_3(t)$

(f) $\widehat{\vartheta}_4(t)$

Figure 3.10 Example 3.4 (SG)

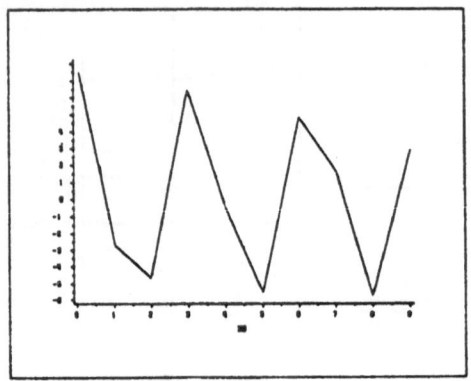

(g) Self-tuning Output Autocorrelation

Figure 3.10   Example 3.4 (SG)

Chapter 4

Adaptive Control of Systems with Random Disturbances

4.1 Introduction

It is not very surprising that the tracking problem has been only modestly extended to the case of LTV systems. The fact is that although adaptive control is an old idea, it has only recently emerged as a branch of Systems and Control Theory with bona fide results. Even for LTI systems, the feedback is non-linear and time varying thereby rendering the analysis of such systems very difficult. It is to be expected that the analysis of LTV systems under adaptive control will be at least as complicated.

In reality, it can be argued that only time varying systems exist and indeed examples such as robotic manipulators, manufacturing processes, and aircraft (see e.g. [3, Åström]) come easily to mind. The kind of time variations or changing dynamics occuring in such systems can be deterministic, stochastic, or a combination of both, and it is the task of an adaptive controller to be alert for such changes. This alertness is reflected in the learning part of the controller which is in turn implemented via a parameter estimator.

This chapter will begin by considering how some of the schemes presented in Chapter 2 perform on LTV systems. Although the studies presented yield only limited results, they are nevertheless an important aid in understanding how robust these algorithms are with respect to changes in the modelled dynamics of the system to be controlled. Ad hoc modifications are then discussed and this is followed by a discussion of the adaptive control of a finite state homogeneous Markov chain. The chapter concludes with the presentation of a new framework for the analysis of a general class of LTV systems with random disturbances following [21, Caines et.al.].

4.2   Robustness of Self-Tuning Controllers

An asymptotically optimal controller for a LTV system should yield results which reduce to corresponding results for a LTI system when the time variations are removed. Having studied time invariant systems, it is natural to ask whether the strategies developed there carry over to LTV systems. It is clear from the outset that if the controller is not able to quickly learn of changes occuring in the system to be controlled, one can rule out all but the very slowest and smallest of time variations in the system dynamics. Along these lines, [2, Anderson and Johnstone] have established robustness results for the adaptive control of deterministic systems (such as those discussed in [42, Goodwin et.al.]) which are based on the exponential convergence of the self-tuning algorithm. The authors argue that exponential stability is a key property for obtaining robust stability results.

In the stochastic environment, [30, Chen and Caines] examined the robustness of the MLS estimation algorithm proposed in [27, Chen] with respect to parameter variations. An earlier study of [16, Caines] established a preliminary result using the SG algorithm. Both these studies simplify to known results for LTI systems when the time variations are removed and are briefly presented below.

Consider SISO LTV ARMAX systems of the form

$$A(t,z)y(t) = B(t,z)u(t-1) + C(t,z)w(t), \qquad t \geq 1, \qquad (4.1a)$$

with initial condition x_0, where

$$A(t,z) = 1 + a_1(t)z + \cdots + a_n(t)z^n, \qquad (4.1b)$$

$$B(t,z) = b_0(t) + b_1(t)z + \cdots + b_n(t)z^n, \quad b_0(t) \neq 0, \qquad (4.1c)$$

and

$$C(t,z) = 1 + c_1(t)z + \cdots + c_n(t)z^n. \qquad (4.1d)$$

The following assumptions are made on (4.1):

(i)   Let $S = \{B(t,z)|B(t,z) = b_0(t) + b_1(t)z + \cdots + b_n(t)z^n, z \in \mathbf{C}, t \geq 1\}$.

Then $\forall B \in S, \quad B(t,z) = 0 \Rightarrow |z| > 1$.

(ii) $x_0 \perp\!\!\!\perp w_1^t \quad \forall t \geq 1$ and $\mathrm{E}\left[\|x_0\|^2\right] < \infty$.

(iii) $\mathrm{E}\left[w(t)|\sigma\{x_0, w_1^{t-1}\}\right] = 0 \qquad \text{a.s.},$

$\mathrm{E}\left[w^2(t)|\sigma\{x_0, w_1^{t-1}\}\right] = \sigma^2 \qquad \text{a.s.},$

$$\limsup_{N \to \infty} \frac{1}{N} \sum_{t=1}^{N} w^2(t) < \infty \qquad \text{a.s.}$$

(iv)   Let $\psi(t) = [a_1(t) \cdots a_n(t), b_0(t) \cdots b_n(t), c_1(t) \cdots c_n(t)]^T$.

Then $\mathrm{E}\left[f(\psi(t+\tau)|\sigma\{x_0, y_1^t, \psi_1^t\}\right] = \mathrm{E}\left[f(\psi(t+\tau))|\sigma\{\psi_1^t\}\right]$

for all measurable $f(\cdot)$ and $\forall \tau \geq 1$.

Note that assumption (i) is the counterpart of (M3) for LTV systems while (iii) is (W1), (W2) and (W5) lumped together. Assumption (iv) simply states that future parameters in (4.1) are independent of past and present inputs and outputs given past and present parameters. Now, in analogy with the developments in Chapter 2, one seeks a predictor model for (4.1). The predictor sought for is the linear least squares predictor $y^\circ(t|t-1)$ of $y(t)$ in the class of predictors measurable with respect to $\sigma\{x_0, y_1^{t-1}, \psi_1^t\}$. It is worth noting that the σ-algebra $\mathcal{F}_{t-1}^y$ of Chapter 2 has been expanded by the addition of ψ_1^t with the result that the parameters enter the predictor in a deterministic way as was the case with the time invariant predictor of Chapter 2. The predictor $y^\circ(t|t-1)$ is given as before by the conditional expectation

$$y^\circ(t|t-1) \stackrel{\Delta}{=} \mathrm{E}\left[y(t)|\sigma\{x_0, y_1^{t-1}, \psi_1^t\}\right] \tag{4.2}$$

and satisfies the equation (see [15, Caines])

$$\overline{C}(t,z)y^\circ(t|t-1) = (\overline{C}(t,z) - A(t,z))y(t) + B(t,z)u(t-1), \qquad t \geq 1, \tag{4.3}$$

where

$$\overline{C}(t,z) = 1 + \overline{c}_1(t)z + \cdots + \overline{c}_n(t)z^n$$

with the coefficients $\overline{c}_i(t)$ given by a Cholesky factorization of the symmetric matrix CC^T where C is the infinite upper triangular banded matrix with $c_i(k)$ in the kth row (counting upwards) and the $(k + 1 - i)^{th}$ column (counting leftwards). The following assumption is made on $\overline{C}(t, z)$:

(v)   For each $t \geq 1$, $\overline{C}(t, z) = 0 \Rightarrow |z| > 1$. Furthermore, $\overline{C}(t, z) - \dfrac{1}{2}$ is uniformly strictly positive real (see [16, Caines]).

Subtracting $\overline{C}(t, z)y^*(t)$ from both sides of (4.3) gives

$$\overline{C}(t, z)(y_u^\circ(t|t - 1) - y^*(t)) = (\overline{C}(t, z) - A(t, z))y(t)$$
$$+ B(t, z)u(t - 1) - \overline{C}(t, z)y^*(t), \quad t \geq 1, \quad (4.4)$$

where as in Chapter 2, y° has been subscripted to emphasize its dependence on u. Letting

$$\boldsymbol{\vartheta}(t) = [(\bar{c}_1(t) - a_1(t)) \cdots (\bar{c}_n(t) - a_n(t)), b_0(t) \cdots b_n(t), \bar{c}_1(t) \cdots \bar{c}_n(t)]^T \quad (4.5a)$$

and

$$\boldsymbol{\varphi}(t) = [y(t) \cdots y(t - n + 1), u(t) \cdots u(t - n), -y^*(t) \cdots - y^*(t - n + 1)]^T \quad (4.5b)$$

equation (4.4) may be written as

$$\overline{C}(t, z)(y_u^\circ(t|t - 1) - y^*(t)) = \boldsymbol{\varphi}^T(t - 1)\boldsymbol{\vartheta}(t) - y^*(t), \quad t \geq 1. \quad (4.5c)$$

This is the time varying version of the predictor model (2.8). Its parameters have a one-to-one relationship with the parameters of (4.1) and are assumed to satisfy

(vi) $E[\boldsymbol{\vartheta}(t + 1)|\sigma\{\boldsymbol{\vartheta}_1^t\}] = \boldsymbol{\vartheta}(t), t \geq 1$ and $\sup\limits_{t \geq 1} E[\|\boldsymbol{\vartheta}(t)\|^2] < \infty$ a.s.

This assumption implies that the martingale $(\boldsymbol{\vartheta}(t), \sigma\{\boldsymbol{\vartheta}_1^t\})$ converges, i.e.

$$\lim_{t \to \infty} \boldsymbol{\vartheta}(t) = \boldsymbol{\vartheta}_\infty \quad \text{a.s.}$$

with $E[\|\boldsymbol{\vartheta}_\infty\|^2] < \infty$. From the analysis of LTI systems, the mean square prediction error $E[v^2(t)|\mathcal{F}_{t-1}^y]$ of (2.7) must now be re-defined since the parameters enter the system and predictor models as random variables. The analogous meaningful quantity is the random variable

$$\gamma^2(\boldsymbol{\vartheta}(t)) \triangleq E[v^2(t)|\sigma\{\boldsymbol{x_0}, y_1^{t-1}, \boldsymbol{\vartheta}_1^t\}]$$

which reduces to the constant γ^2 of (2.7) in the case of constant parameters. Let

$$\pi_N \triangleq \frac{1}{N}\sum_{t=1}^{N} E[\gamma^2(\vartheta(t))|\sigma\{\vartheta_1^{t-1}\}]$$

and assume

(vii) $\quad \sup_{t\geq1} E[\gamma^2(\vartheta(t))|\sigma\{\vartheta_1^{t-1}\}] < \infty$ a.s., and $\quad \pi_\infty \triangleq \lim_{N\to\infty} \pi_N$ exists

and is finite a.s.

Then, as in Chapter 2, using the self-tuning approach, consider the estimation of ϑ and its use in the control strategy (2.37a). Under the assumption

$(viii)$ $\quad$ The random variables $\quad x_0, w_1^t, \vartheta_1^t \quad$ are jointly absolutely continuous

w.r.t. Lebesgue measure for $\quad t \geq 1$,

the following theorem (compare with theorem 2.5) establishes an intuitively plausible result.

Theorem 4.1 [16, Caines] (SG)

Consider the tracking problem for the system (4.1) using algorithm (2.27) with (4.5) and the adaptive control (2.37a). Let assumptions (i) - (viii) hold. Then

$$\limsup_{N\to\infty}\frac{1}{N}\sum_{t=1}^{N} y^2(t) < \infty \qquad \text{a.s.} \tag{4.6a}$$

$$\limsup_{N\to\infty}\frac{1}{N}\sum_{t=1}^{N} u^2(t) < \infty \qquad \text{a.s.} \tag{4.6b}$$

and

$$\lim_{N\to\infty}\frac{1}{N}\sum_{t=1}^{N} E[(y(t)-y^*(t))^2|\sigma\{x_0,y_1^{t-1},\vartheta_1^t\}] = \pi_\infty \quad \text{a.s.} \tag{4.6c}$$

■

Random time variations in the parameters of the converging martingale type described above are difficult to find in practical applications. A more plausible type of random variation is described in [30, 28, Chen and Caines] and their main results are presented in theorem 4.2 below.

Consider the ARX system model

$$A(t,z)y(t) = B(t,z)u(t-1) + w(t), \qquad t \geq 1, \qquad (4.7a)$$

with initial condition x_0 and where

$$A(t,z) = 1 + a_1(t)z + \cdots + a_n(t)z^n, \qquad n \geq 1, \qquad (4.7b)$$

$$B(t,z) = b_0(t) + b_1(t)z + \cdots + b_m(t)z^m, \qquad b_0(t) \neq 0,\ m \geq 0. \qquad (4.7c)$$

The time variations in the parameters $a_i(t)$ of (4.7) are given by

$$a_i(t) = a_i + v_i(t), \qquad i = 1, \cdots, n \qquad (4.8a)$$

$$b_i(t) = b_i + \varsigma_i(t), \qquad i = 0, \cdots, m \qquad (4.8b)$$

where a_i and b_i are constant and unknown, while $v_i(t)$ and $\varsigma_i(t)$ are random variables satisfying

$$E[v_i(t)|\mathcal{F}_{t-1}] = 0, \quad E[\varsigma_i(t)|\mathcal{F}_{t-1}] = 0 \quad \text{a.s.} \qquad (4.9a)$$

$$E[v_i^2(t)|\mathcal{F}_{t-1}] < \infty, \quad E[\varsigma_i^2(t)|\mathcal{F}_{t-1}] < \infty \quad \text{a.s.} \qquad (4.9b)$$

$$E[v_i(t)v_j(t)|\mathcal{F}_{t-1}] = 0 \quad i \neq j, \ E[\varsigma_i(t)\varsigma_j(t)|\mathcal{F}_{t-1}] = 0 \quad i \neq j, \text{ a.s.} \qquad (4.9c)$$

$$|v_i(t)| < \alpha, \quad |\varsigma_i(t)| < \beta, \quad t \geq 1, \qquad (4.9d)$$

where α and β are specified below and $\mathcal{F}_t = \sigma\{x_0, w_1^t, v_1^t, \varsigma_1^t\}$. The disturbance process $\{w(t)\}$ satisfies

$$E[w(t)|\mathcal{F}_{t-1}] = 0 \qquad \text{a.s.} \qquad (4.9e)$$

$$E[w^2(t)|\mathcal{F}_{t-1}] < \infty \qquad \text{a.s.} \qquad (4.9f)$$

$$E[w^2(t)|\mathcal{F}_{t-1}] \leq \beta_2 r^{\delta_2}(t) \quad \text{a.s.} \quad \beta_2 > 0,\ \delta_2 \in [0,1) \qquad (4.9g)$$

$$w^2(t) \leq \beta_1 r^{\delta_1}(t) \text{a.s.} \quad \beta_1 > 0, \delta_1 \in [0,1) \qquad (4.9h)$$

$$\lim_{N \to \infty} \frac{1}{N} \sum_{t=1}^{N} w^2(t) = \sigma_w^2(\omega) > 0 \qquad \text{a.s.} \qquad (4.9i)$$

where $\{r(t)\}$ is a scalar random sequence tending to infinity in t and is defined below. It is also assumed that for $i = 1, \cdots, n, j = 0, \cdots m$,

$$E[w(t)v_i(t)|\mathcal{F}_{t-1}] = 0, \quad E[w(t)\varsigma_j(t)|\mathcal{F}_{t-1}] = 0 \quad \text{a.s.} \qquad (4.9j)$$

and that $B(z) = b_0 + b_i z + \cdots + b_m z$ has all its roots strictly outside the unit circle. To rewrite (4.7a) in the usual vector notation let

$$\vartheta = [-a_1 \cdots - a_n, b_0 \cdots b_m]^T, \qquad (4.10a)$$

$$\varphi(t) = [y(t) \cdots y(t - n + 1), u(t) \cdots u(t - m)]^T, \qquad (4.10b)$$

whence (4.7a) becomes

$$y(t) = \varphi^T(t - 1)\vartheta + w(t) + \varepsilon(t),$$

$$\varepsilon(t) = -\sum_{i=1}^{n} v_i(t)y(t - i) + \sum_{j=0}^{m} \varsigma_j(t)u(t - j).$$

The process $\{\varepsilon(t)\}$ summarizes the effect of the random parameter variations while ϑ contains their mean values.

Consider now the estimation of ϑ via the MLS algorithm (2.31) with Modification 1 and its use in the control strategy (2.37a). The sequence $\{r(t)\}$ in (4.9g,h) is as defined in Modification 1 while α and β of (4.9d) are chosen so that

$$1 - \alpha(1 + (n + m)K_1)(z + \cdots + z^n) = 0 \;\Rightarrow\; |z| > 1, \qquad (4.11a)$$

and

$$B(t, z) = 0 \Rightarrow |z| > 1 \quad \forall t \geq 1. \qquad (4.11b)$$

Condition (4.11a) will be ensured if

$$\alpha < 1/((1 + (n + m)K_1)n). \qquad (4.11c)$$

Then, the following important result is established.

Theorem 4.2 [30, 28, Chen and Caines] (MLS)

Consider the tracking problem for the system (4.7) using algorithm (2.31) with Modification 1 and (4.10).

(i) Let assumptions (4.9) hold with $\varsigma_i(t) \equiv 0, i = 0, \cdots, m$, and with α such that (4.11a) is satisfied. Then (4.6a,b) hold while (4.6c) is replaced by

$$\limsup_{N \to \infty} \frac{1}{N} \sum_{t=1}^{N} (y(t) - y^*(t))^2 = \sigma_w^2 + \limsup_{N \to \infty} \frac{1}{N} \sum_{t=1}^{N} \varepsilon^2(t). \qquad (4.12a)$$

Further, there exists $\rho \in (0,1)$ such that

$$\limsup_{N\to\infty} \frac{1}{N} \sum_{t=1}^{N} \epsilon^2(t) \leq \frac{6\alpha^2 n}{(1-\rho)^2}\left((1+(n+m)K_1)^2 \sigma_w^2 + \sup_{t\geq 1} |y^*(t)|^2 \right). \qquad (4.12b)$$

(ii) Further to (i) above restrict the control $u(t)$ so that

$$|u(t)|^2 \leq cr^\epsilon(t), \quad c > 0, \quad 0 \leq \epsilon < 1, \qquad (4.12c)$$

and let $\varsigma_i(t) \not\equiv 0$. Then the result (4.12a) holds.

∎

Corollary 4.2a [28, Chen and Caines]

In both cases of the above theorem, if $A(t,z)$ has all its roots strictly outside the unit circle for all $t \geq 1$ with $\sum_{i=1}^{t} \varphi(i)\varphi^T(i) + (1/(n+m))I$ having a finite ratio of maximum to minimum eigenvalues, then

$$\lim_{t\to\infty} \widehat{\vartheta}(t) = \vartheta \qquad \text{a.s.} \qquad (4.12d)$$

∎

Note that the right hand side of (4.12b) goes to zero as $\alpha \to 0$.

4.3   From Self-Tuning Controllers to Adaptive Controllers

The self-tuning control algorithms described in Chapter 2 use parameter estimators whose gain is an a.s. strictly decreasing sequence. The structure of the estimation algorithms under consideration has the general form

$$\widehat{\vartheta}(t) = \widehat{\vartheta}(t-1) + \gamma(t)\mathbf{P}(t)\boldsymbol{\eta}(t)(y(t) - \widehat{y}(t))$$

where $\widehat{\vartheta}(t)$ is the parameter estimate at time t, $\{\gamma(t)\}$ is a sequence of scalars usually referred to as the gain sequence, $\mathbf{P}(t)$ a matrix computed from observations up to time t, and $\widehat{y}(t)$ a prediction of the output $y(t)$ based on measurements up to time $t-1$. The vector $\boldsymbol{\eta}(t)$ is constructed from previous observations and is usually related to the gradient of $\widehat{y}(t)$ w.r.t. ϑ (see [56, Ljung and Söderström], [43, Goodwin and Sin], [15, Caines]).

In all the algorithms presented, the two kinds of gain sequences encountered are the scalar sequence $\{r^{-1}(t)\}$ where

$$r(t) = \sum_{i=1}^{t} \varphi^T(i)\varphi(i), \tag{4.13}$$

and the matrix sequence $\{\mathbf{P}^{-1}(t)\}$ where

$$\mathbf{P}(t) = \sum_{i=1}^{t} \varphi(i)\varphi^T(i). \tag{4.14}$$

Such sequences pay equal importance to the data in the sense that they weight all the observations equally. At each time instant, a non-negative contribution is made to $\{r(t)\}$ and $\{\mathbf{P}(t)\}$ so that new observations have less and less relative strength resulting in a controller which is less and less sensitive to prediction errors. Such behaviour is clearly undesirable for tracking time varying parameters and reflects the fact that self-tuning controllers need not — and in fact do not — maintain the alertness required of an adaptive controller. The strategy then is clear: prevent the gain sequence from decreasing to zero so that the controller will be sensitive to changing dynamics. Caution should be exercised however not to allow the gain sequence to become too large since this would result in tracking the noise variations.

A popular method implementing the aforementioned strategy is one which uses a forgetting factor. This refers to a sequence of scalars $\{\lambda(t,i)\}$ which unbalance the uniform observation weighting profile in (4.13) and (4.14) in favour of more recent observations. The latter equations become respectively

$$r(t) = \sum_{i=1}^{t} \lambda(t,i)\varphi^T(i)\varphi(i) \tag{4.15}$$

and

$$\mathbf{P}(t) = \sum_{i=1}^{t} \lambda(t,i)\varphi(i)\varphi^T(i) \tag{4.16}$$

where $0 \le \lambda(t,i) \le 1,\quad \forall t \ge 1,\quad 1 \le i \le t.$ The case $\lambda(t,i) = \lambda^{t-i},\quad 0 < \lambda < 1,$ corresponds to geometrically discounting old data in which case the sequence $\{r(t)\}$ can be recursively computed via

$$r(t) = \lambda r(t-1) + \varphi^T(t)\varphi(t) \tag{4.17}$$

while $\{\mathbf{P}^{-1}(t)\} \underset{\nabla}{=} \{\mathbf{R}(t)\}$ via

$$\mathbf{R}(t) = \frac{1}{\lambda}\left(\mathbf{R}(t-1) - \frac{\mathbf{R}(t-1)\boldsymbol{\varphi}(t)\boldsymbol{\varphi}^T(t)\mathbf{R}(t-1)}{\lambda + \boldsymbol{\varphi}^T(t)\mathbf{R}(t-1)\boldsymbol{\varphi}(t)}\right). \qquad (4.18)$$

There are some obvious dangers with the above procedure. If there are lengthy periods of little or no new information (reflected in small prediction errors), the gain sequence can become quite large at a rate which depends on the choice of λ. This can cause temporary instabilities due to the extreme sensitivity attained or simply due to numerical errors (see [39, Fortescue et.al.]). The situation can be remedied by considering a time varying λ in (4.17) and (4.18). This variation should depend on the current performance of the algorithm in a manner whereby large prediction errors, possibly signaling changes in system dynamics, should dictate a λ which forgets old data, while small prediction errors should dictate a λ closer to unity. A scheme with this underlying idea is proposed in [39, Fortescue et.al.] where the self-tuning regulator of [5, Åström and Wittenmark] using RLS estimation is considered. The choice of λ is such that the sum of squares of the a posteriori prediction errors is kept at a user specified level Σ_0. The rule is to compute $\lambda(t)$ via

$$\lambda(t) = 1 - \frac{1}{N(t)},$$

$$N(t) = \frac{\Sigma_0(\boldsymbol{\varphi}^T(t-1)\mathbf{R}(t-2)\boldsymbol{\varphi}(t-1) + \lambda(t-1))}{\lambda(t-1)(y(t) - \boldsymbol{\varphi}^T(t-1)\hat{\boldsymbol{\vartheta}}(t-1))^2}.$$

To simplify numerical computation, the following approximation of $N(t)$ avoids solving a quadratic for $\lambda(t)$ and is claimed in [39, Fortescue et. al.] to make a small practical difference in most cases, provided $\lambda(t)$ is not allowed to fall below some minimum value $\lambda_{\min}$. Set

$$N(t) = \frac{\Sigma_0(\boldsymbol{\varphi}^T(t-1)\mathbf{R}(t-2)\boldsymbol{\varphi}(t-1))}{(y(t) - \boldsymbol{\varphi}^T(t-1)\hat{\boldsymbol{\vartheta}}(t-1))^2}$$

and compute $\mathbf{R}(t)$ via

$$\mathbf{R}(t-1) = \frac{1}{\lambda(t)}\left(\mathbf{R}(t-2) - \frac{\mathbf{R}(t-2)\boldsymbol{\varphi}(t-1)\boldsymbol{\varphi}^T(t-1)\mathbf{R}(t-2)}{1 + \boldsymbol{\varphi}^T(t-1)\mathbf{R}(t-2)\boldsymbol{\varphi}(t-1)}\right).$$

The successful implementation of this scheme on a large scale chemical plant is reported in [52, Kershenbaum and Fortescue] and its convergence properties in a deterministic setting are found in [33, Cordero and Mayne]. A different approach is taken in [46, Hägglund] where the strategy adopted is to discount only incoming data so that a constant amount of information is retained. Information is represented by the inverse covariance of the parameter prediction error, $\mathbf{P}^{-1}(t)$, and is updated via

$$\mathbf{P}^{-1}(t) = \mathbf{P}^{-1}(t-1) + \left(\frac{1}{v(t)} - \alpha(t) \right) \boldsymbol{\varphi}(t) \boldsymbol{\varphi}^T(t)$$

where $v(t)$ is a measurement variance (see [46, Hägglund]). The positive scalar $\alpha(t)$, for which upper bounds are given, determines how much information is removed at time t.

Apart from the forgetting factor strategy there are various other methods of preventing the gain of the parameter estimator from decaying to zero. One may take the case of forgetting factors to the limit and only consider a fixed number N of previous observations. Also possible is the periodic resetting of the gain to some suitable value or its modification at each iteration by the addition of a suitable quantity. These methods are described in [43, Goodwin and Sin] where the preferred method is said to be the gain resetting method. A similar conclusion was reached in [45, Goodwin and Teoh] where deterministic time varying systems were examined. In fact, the latter study suggests the use of gain resetting even for time invariant systems. It must be remarked that, due to their complexity, it is extremely difficult to mathematically analyse the aforementioned ingenious schemes even though they are effective in simulation and practice.

4.4   Adaptive Control of a Homogeneous Markov Chain

Naturally, to obtain meaningful results for LTV systems, one has to narrow down the kind of parameter variations allowed. For example, in section 4.2 the artificial situation of a converging martingale parameter process was examined and was shown to yield satisfactory results. Also, the more realistic assumption of a randomly time varying bounded parameter process was presented with another set of optimality results. In this section, the assumption that the parameter process is a finite state homogeneous Markov chain will be adopted. The purpose of this section is to demonstrate the derivation of a truly optimal adaptive controller, i.e. one that does not rely on the certainty equivalence principle but rather generates optimal control actions at each step.

Consider the TV ARX model

$$y(t) = \varphi^T(t-1)\vartheta(t) + w(t), \quad t \geq 1, \qquad (4.19a)$$

with initial condition x_0 where

$$\vartheta(t) = [-a_1(t) \cdots - a_n(t), b_0(t) \cdots b_m(t)]^T \qquad (4.19b)$$

and

$$\varphi(t) = [y(t) \cdots y(t-n+1), u(t) \cdots u(t-m)]^T. \qquad (4.19c)$$

Let

$$\mathcal{G}_t = \sigma\{x_0, y(1) \cdots y(t), \vartheta(1) \cdots \vartheta(t)\}, \quad t \geq 1, \qquad (4.20a)$$

$$\mathcal{F}_t = \sigma\{x_0, y(1) \cdots y(t)\}, \qquad t \geq 1, \qquad (4.20b)$$

$$\mathcal{F}_0 = \mathcal{G}_0 \overset{\triangle}{=} \sigma\{x_0\}, \qquad (4.20c)$$

and note that via (4.19) and provided $u_t \in \mathcal{F}_t$,

$$\mathcal{H}_t \overset{\triangle}{=} \sigma\{x_0, w(1) \cdots w(t), \vartheta(1) \cdots \vartheta(t)\} = \mathcal{G}_t, \quad t \geq 1.$$

Assuming that $(w(t), \sigma\{x_0, w(1) \cdots w(t)\})$ is a martingale difference process it then follows that $(w(t), \mathcal{F}_t)$ is also such a process, i.e.

$$E[w(t)|\mathcal{F}_{t-1}] = 0 \qquad \forall t \geq 1. \qquad (4.20d)$$

Further assume that $\{\vartheta(t)\}$ is an unobserved process with

$$\sigma\{\vartheta(1) \cdots \vartheta(t)\} \perp\!\!\!\perp \sigma\{x_0, w(1) \cdots w(t)\}, \qquad \forall t \geq 1. \qquad (4.20e)$$

A control u_{mv} is desired which will minimize the variance of y. It is well known that

$$y^\circ(t|t-1) \overset{\triangle}{=} E[y(t)|\mathcal{F}_{t-1}]$$

is the unique (up to sets of $\mathcal{F}_{t-1}$-conditional probability zero) $\mathcal{F}_{t-1}$ measurable least squares estimate of $y(t)$. Using (4.19) and (4.20d)

$$y_u^\circ(t|t-1) = \varphi^T(t-1)\vartheta^\circ(t|t-1)$$

where $\vartheta^\circ(t|t-1) = \mathrm{E}\{\vartheta(t)|\mathcal{F}_{t-1}\}$, and as before the subscript u of y^0 indicates its dependence on u. Then, choosing $u_{mv}(t-1)$ the $\mathcal{F}_{t-1}$ measurable random variable solving

$$\varphi^T(t-1)\vartheta^\circ(t|t-1) = 0, \tag{4.21}$$

will minimize $\mathrm{E}[y^2(t)]$, i.e. will yield a minimum variance control. In contrast with the previous sections, the process $\{\vartheta(t)\}$ will be restricted to lie in a finite set via the following assumptions:

$$\vartheta(t) : \Omega \to S = \{\mathbf{s}_1 \cdots \mathbf{s}_p\}, \qquad \mathbf{s}_i \in \Re^{n+m+1}, \qquad 1 \le i \le p, \tag{4.22a}$$

$$\mathrm{E}[\vartheta(t)|\sigma\{\vartheta(1)\cdots\vartheta(t-1)\}] = \mathrm{E}[\vartheta(t)|\sigma\{\vartheta(t-1)\}], \tag{4.22b}$$

$$(\mathbf{P_T})_{ij} \overset{\Delta}{=} P(\vartheta(t) = \mathbf{s}_j|\vartheta(t-1) = \mathbf{s}_i), \quad 1 \le i,j \le p. \tag{4.22c}$$

Note that the transition probabilites in (4.22c) are independent of t and that P is a discrete conditional probability density function. The conditional expectation $\vartheta^\circ(t|t-1)$ is then given by [*]

$$\vartheta^\circ(t|t-1) = \sum_{i=1}^p \mathbf{s}_i z_i(t-1) \tag{4.23}$$

where for each $t \ge 1, z_i(t-1) \overset{\Delta}{=} P(\vartheta(t) = \mathbf{s}_i|\mathcal{F}_{t-1})$ is a random variable which can be written as

$$z_i(t-1) = \mathrm{E}[\mathbb{1}_{\vartheta(t)=\mathbf{s}_i}(\omega)|\mathcal{F}_{t-1}].$$

Then, since $\mathcal{F}_{t-1} \subset \mathcal{G}_{t-1}$,

$$z_i(t-1) = \mathrm{E}[\mathrm{E}[\mathbb{1}_{\vartheta(t)=\mathbf{s}_i}(\omega)|\mathcal{G}_{t-1}]|\mathcal{F}_{t-1}]$$

$$= \mathrm{E}[\mathrm{E}[\mathbb{1}_{\vartheta(t)=\mathbf{s}_i}(\omega)|\sigma\{\vartheta_1^{t-1}\}]|\mathcal{F}_{t-1}]$$

where (4.20e) was used to obtain the last equality. The Markov property (4.22b) then gives

$$z_i(t-1) = \mathrm{E}[\mathrm{E}[\mathbb{1}_{\vartheta(t)=\mathbf{s}_i}(\omega)|\sigma\{\vartheta(t-1)\}]|\mathcal{F}_{t-1}]$$

$$= \sum_{j=1}^p (\mathbf{P_T})_{ji} P(\vartheta(t-1) = \mathbf{s}_j|\mathcal{F}_{t-1}). \tag{4.24}$$

---

[*] Conversations with Sean Meyn are gratefully acknowledged for the following analysis.

To obtain an expression for the discrete conditional density in the above equation, let $y(t)$ have a conditional density function $f_y(\cdot|\mathcal{F}_{t-1})$ and write

$$P(\vartheta(t-1) = s_j|\mathcal{F}_{t-1}) = P(\vartheta(t-1) = s_j|y(t-1), \mathcal{F}_{t-2}).$$

Then, by the definition of a conditional probability density,

$$\int_A P(\vartheta(t-1) = s_j|y(t-1), \mathcal{F}_{t-2}) f_y(y(t-1)|\mathcal{F}_{t-2}) dy =$$

$$\int_A f_y(y(t-1)|\vartheta(t-1) = s_j, \mathcal{F}_{t-2}) P(\vartheta(t-1) = s_j|\mathcal{F}_{t-2}) dy$$

up to sets of zero measure with respect to the conditional distribution $P(\vartheta(t-1), y(t-1)|\mathcal{F}_{t-2})$. For the desired expression, the above equation is rewritten as an equation in conditional densities as

$$P(\vartheta(t-1) = s_j|\mathcal{F}_{t-1}) = \frac{f_y(y(t-1)|\vartheta(t-1) = s_j, \mathcal{F}_{t-2}) P(\vartheta(t-1) = s_j|\mathcal{F}_{t-2})}{f_y(y(t-1)|\mathcal{F}_{t-2})} \quad (4.25)$$

which can be seen to be a mixed discrete and continuous Bayes formula. The conditional density in the denominator of (4.25) is given by

$$f_y(y(t-1)|\mathcal{F}_{t-2}) = \sum_{j=1}^{p} f_y(y(t-1)|\vartheta(t-1) = s_j, \mathcal{F}_{t-2}) P(\vartheta(t-1) = s_j|\mathcal{F}_{t-2}). \quad (4.26)$$

Adopting the inductive assumption that $P(\vartheta(t-1) = s_j|\mathcal{F}_{t-2})$ is available at the instant $t = 1$ for all values $s_j, j = 1 \cdots p$, one is left with finding an expression for f_y on the right hand side of (4.26). Using (4.19),

$$f_y(y(t-1)|\vartheta(t-1) = s_j, \mathcal{F}_{t-2}) = f_y(\varphi^T(t-2)\vartheta(t-1) + w(t-1)|\vartheta(t-1) = s_j, \mathcal{F}_{t-2})$$

$$= f_y(\varphi^T(t-2)s_j + w(t-1))$$

via (4.20e) and the $\mathcal{F}_{t-2}$ measurability of $\varphi(t-2)$. Then, by a simple change of variables

$$f_y(y(t-1)|\vartheta(t-1) = s_j, \mathcal{F}_{t-2}) = f_w(y(t-1) - \varphi^T(t-2)s_j) \quad (4.27)$$

where f_w is the density of the $\{w(t)\}$ process. Combining (4.25) - (4.27) gives

$$P(\vartheta(t-1) = s_j|\mathcal{F}_{t-1}) = \frac{f_w(y(t-1) - \varphi^T(t-2)s_j) P(\vartheta(t-1) = s_j|\mathcal{F}_{t-2})}{\sum_{i=1}^{p} f_w(y(t-1) - \varphi^T(t-2)s_i) P(\vartheta(t-1) = s_i|\mathcal{F}_{t-2})}.$$

$$(4.28)$$

Finally, (4.23), (4.24), and (4.28) give the desired expression for $\vartheta^\circ(t|t-1)$,

$$\vartheta^\circ(t|t-1) = \sum_{i=1}^{p} \mathbf{s}_i \left(\sum_{j=1}^{p} (\mathbf{P_T})_{ji} \frac{f_w(y(t-1) - \varphi^T(t-2)\mathbf{s}_j) P(\vartheta(t-1) = \mathbf{s}_j|\mathcal{F}_{t-2})}{\sum_{k=1}^{p} f_w(y(t-1) - \varphi^T(t-2)\mathbf{s}_k) P(\vartheta(t-1) = \mathbf{s}_k|\mathcal{F}_{t-2})} \right).$$

(4.29)

The control u_{mv} given by (4.21) and (4.29) may be computed provided the set $S = \{\mathbf{s}_1 \cdots \mathbf{s}_p\}$ and the associated transition probability matrix $\mathbf{P_T}$ are known, as well as the distribution of the $\{w(t)\}$ process. Such a general result is not available if the state space of the $\{\vartheta(t)\}$ process is uncountable and, as has been the case so far, various approximate estimation algorithms for $\{\vartheta(t)\}$ were used instead. An exception to this is when $\{w(t)\}$ is normally distributed whence the Kalman filter can be used to generate the desired conditional expectation. This case yields some promising results as shown in the next section. An alternative for the calculation of the control law is to use the maximum likelihood estimate $\vartheta^*(t|t-1)$ of $\vartheta(t)$ defined by

$$\vartheta^*(t|t-1) \triangleq \arg\max_{\mathbf{s}_i \epsilon S} P(\vartheta(t) = \mathbf{s}_i|\mathcal{F}_{t-1})$$

and given by

$$\vartheta^*(t|t-1) = \arg\max_{\mathbf{s}_i \epsilon S} \sum_{j=1}^{p} (\mathbf{P_T})_{ji} \frac{f_w(y(t-1) - \varphi^T(t-2)\mathbf{s}_j) P(\vartheta(t-1) = \mathbf{s}_j|\mathcal{F}_{t-2})}{\sum_{k=1}^{p} f_w(y(t-1) - \varphi^T(t-2)\mathbf{s}_k) P(\vartheta(t-1) = \mathbf{s}_k|\mathcal{F}_{t-2})}.$$

(4.30)

4.5 Stochastic Stability and Adaptive Control of TV ARMAX Systems

A quite general model for time vaying parameters is that of a Borel function F of a Markov process $\{\mathbf{x}(t)\}$, i.e.

$$\vartheta(t) = F(\mathbf{x}(t)).$$

(4.31)

For example, the ARMA (n,m) model

$$\vartheta(t) + \mathbf{A}_1\vartheta(t-1) + \cdots + \mathbf{A}_n\vartheta(t-n) = \mathbf{w}(t) + \mathbf{B}_1\mathbf{w}(t-1) + \cdots + \mathbf{B}_m\mathbf{w}(t-m), \quad t \geq 1,$$

where $\{\mathbf{w}(t)\}$ is an i.i.d. vector process, is a special case of (4.31).

Letting

$$\mathbf{x}(t) = [\vartheta^T(t) \cdots \vartheta^T(t-n), \mathbf{w}^T(t) \cdots \mathbf{w}^T(t-m)]^T$$

one can translate equation (4.31) into

$$\mathbf{x}(t+1) = \mathbf{T}\mathbf{x}(t) + \boldsymbol{\eta}(t+1) \tag{4.32}$$

for some constant coefficient matrix $\mathbf{T}$ and an i.i.d. process $\{\boldsymbol{\eta}(t)\}$ whence the process $\{\mathbf{x}(t)\}$ is seen to be Markov. As will be seen from a simple example that follows, the state vector $\mathbf{x}(t)$ may be enlarged by parameter estimates, observations, and other related quantities and the powerful theory of Markov processes can then be used to yield important new results. The main references for the discussion which follows are [21, Caines et.al.] and [58, Meyn and Caines]. Note that while the discussion takes place in the setting of $\Re^3$ and its Borel sets $\mathcal{B}^3$, the results presented are valid for a subset S of the general multidimensional Euclidean space $\Re^n$ and its Borel sets $\mathcal{B}(S)$.

The following simple example is considered. Let

$$\vartheta(t+1) = \alpha\vartheta(t) + e(t), \quad |\alpha| < 1, \quad t \geq 1, \tag{4.33a}$$

$$y(t) = \vartheta(t)y(t-1) + u(t-1) + w(t), \qquad t \geq 1, \tag{4.33b}$$

where the $\{\vartheta(t)\}$ process is assumed unknown. Define the σ-algebra of observations

$$\mathcal{F}_t^y \triangleq \sigma\{\vartheta(1), u(0), y(0) \cdots y(t)\} \underset{\triangledown}{=} \sigma\{\mathbf{x_0}, y(1) \cdots y(t)\} \tag{4.34}$$

and assume the following:

(i) $\{e(t)\}$ and $\{w(t)\}$ are scalar, Gaussian, zero mean processes with

$$\mathbf{E}\left[\begin{bmatrix} e(t) \\ w(t) \end{bmatrix} [e(s)w(s)]\right] = \begin{bmatrix} \sigma_e^2 & 0 \\ 0 & \sigma_w^2 \end{bmatrix} \delta(t-s), \quad \forall t, s \in \mathcal{N},$$

where $\delta(\cdot)$ is the Kronecker δ.

(ii) $\mathbf{x_0} = [\vartheta(1), u(0), y(0)]^T$ is normally distributed with

$$\mathbf{E}[\mathbf{x_0}[e(t)w(t)]] = \mathbf{0},$$

where $\mathbf{0}$ is the 3×2 zero matrix.

These assumptions imply that $\{w(t)\} \perp\!\!\!\perp \{e(t)\}$, $\{w(t)\} \perp\!\!\!\perp \{\vartheta(t)\}$, and via equation (4.33), that $\{w(t)\}$ and $\{e(t)\}$ are independent of $\mathcal{F}_{t-1}^y$.

The problem under consideration is that of regulation about zero, i.e. tracking the sequence $y^*(t) = 0$. As was the case until now, following the self-tuning methodology, the control is determined by setting the best (MV) prediction of the output to $y^*(t)$ where this prediction is determined by substituting parameter estimates in place of the actual predictor parameters. The control $u(t)$ is (as always) required to be $\mathcal{F}_t^y$-measurable. It has already been established that the unique $\mathcal{F}_{t-1}^y$-measurable least squares estimate of $y(t)$ is given by

$$y^\circ(t|t-1) = \mathrm{E}[y(t)|\mathcal{F}_{t-1}^y] \tag{4.35a}$$

where, using (4.33b) and assumptions (i) and (ii),

$$y^\circ(t|t-1) = \mathrm{E}[\vartheta(t)|\mathcal{F}_{t-1}^y]y(t-1) + u(t-1). \tag{4.35b}$$

Further, it is well known that, under the aforementioned assumptions, $\mathrm{E}[\vartheta(t)|\mathcal{F}_{t-1}^y]$ — and hence $y^\circ(t|t-1)$ — can be obtained via the Kalman filter. Then, setting $y^\circ(t|t-1)$ to $y^*(t)$ (i.e. to zero) gives the control strategy

$$u(t) = -\mathrm{E}[\vartheta(t+1)|\mathcal{F}_t^y]y(t), \qquad t \geq 1. \tag{4.36}$$

To simplify the analysis, the unity coefficient of $u(t-1)$ in (4.33b) will be assumed known (i.e. u is permitted to be a function of this quantity and its value here is 1) and a 1-dimensional Kalman filter will be used. For the state space model

$$x(t+1) = f(t)x(t) + g(t)u(t) + e(t)$$
$$z(t) = h(t)x(t) + w(t)$$

one makes the following correspondences with reference to (4.33):

$$z(t) \leftrightarrow y(t) - u(t-1), \qquad u(t) \leftrightarrow u(t), \qquad x(t) \leftrightarrow \vartheta(t),$$
$$f(t) \leftrightarrow \alpha, \qquad g(t) \leftrightarrow 0, \qquad h(t) \leftrightarrow y(t-1). \tag{4.37}$$

Then one gets the state space model

$$\vartheta(t+1) = \alpha\vartheta(t) + e(t) \tag{4.38a}$$

$$z(t) = \vartheta(t)y(t-1) + w(t) \tag{4.38b}$$

with the corresponding filter equations

$$\vartheta^\circ(t+1|t) = \alpha\vartheta^\circ(t|t-1) + \frac{\alpha V(t|t-1)y(t-1)}{y^2(t-1)V(t|t-1) + \sigma_w^2} \times$$
$$(z(t) - \vartheta^\circ(t|t-1)y(t-1)), \quad t \geq 1, \qquad (4.39a)$$

$$V(t+1|t) = \frac{\alpha^2 V(t|t-1)\sigma_w^2}{y^2(t-1)V(t|t-1) + \sigma_w^2} + \sigma_e^2, \qquad t \geq 1, \qquad (4.39b)$$

where

$$\vartheta^\circ(t+1|t) = \mathrm{E}\left[\vartheta(t+1)|\mathcal{F}_t^y\right]$$

and

$$V(t+1|t) = \mathrm{E}\left[(\vartheta(t+1) - \vartheta^\circ(t+1|t))^2 \,|\mathcal{F}_t^y\right]$$

provided $\vartheta^\circ(1|0) = \mathrm{E}[\vartheta(1)]$ and $V(1|0) = \mathrm{E}[(\vartheta(1) - \vartheta^\circ(1|0))^2] = 0$.

Using the control strategy (4.36), the closed loop system is given by

$$y(t) = \tilde{\vartheta}(t)y(t-1) + w(t), \qquad t \geq 1, \qquad (4.40a)$$

where

$$\tilde{\vartheta}(t) \triangleq \vartheta(t) - \vartheta^\circ(t|t-1). \qquad (4.40b)$$

Then, using (4.33a), (4.36), (4.37), (4.39a), and (4.40), one gets for $\tilde{\vartheta}$

$$\tilde{\vartheta}(t+1) = \alpha\tilde{\vartheta}(t) - \frac{\alpha V(t|t-1)y(t-1)(\tilde{\vartheta}(t)y(t-1) + w(t))}{y^2(t-1)V(t|t-1) + \sigma_w^2} + e(t+1), \qquad t \geq 1. \quad (4.41)$$

Finally, lumping (4.39b), (4.40a), and (4.41) into one vector gives

$$\mathbf{x}(t) \triangleq \begin{bmatrix} V(t+1|t) \\ \tilde{\vartheta}(t+1|t) \\ y(t) \end{bmatrix} = F(\mathbf{x}(t-1), \boldsymbol{\eta}(t)), \qquad t \geq 1, \qquad (4.42)$$

where $F : \Re^5 \to \Re^3$ is continuous and $\{\boldsymbol{\eta}(t)\}$ is an i.i.d. process in $\Re^2$.
It may be verified that:

(a) $\{\mathbf{x}(t)\}$ is a Markov process with state space $\Re^+ \times \Re^2$ and stationary transition probabilities.

(b) The stochastic transition function $p : \Re^3 \times \mathcal{B}^3 \to [0,1]$ (see Appendix C for definition) of the process $\{\mathbf{x}(t)\}$ satisfies the *Feller property*: for any bounded continuous $f : \Re^3 \to \Re$, $\int f(\mathbf{x})p(\mathbf{s}, d\mathbf{x})$ is continuous in s and hence uniformly continuous on compact subsets of $\Re^3$.

(c) The sets

$$\mathcal{F}_\nu^I \triangleq \{A \in B^3 | p(\xi, A) = \mathbb{1}_A(\xi) \text{ for a.a.}(\nu) \ \xi \in \Re^3\}$$
$$\mathcal{F}_\nu^D \triangleq \{A \in B^3 | p(\xi, A) = \mathbb{1}_B(\xi) \text{ for a.a.}(\nu) \ \xi \in \Re^3 \text{ and some } B \in B^3\}$$

each consist of $\Re^3$ and the null set.

One can think of the evolution of the distributions or measures $\{\mu_t\}$ much like the evolution of the state $\mathbf{x}(t)$ in a state space system. This is elegantly described in the framework of semidynamical systems (see Appendix C for definition). Briefly, consider the set $\mathcal{M}$ of all probability measures μ whose domain is the set of Borel subsets (of $\Re^3$ in this case), and for each $t \geq 1$, define the operator $T(t) : \mathcal{M} \to \mathcal{M}$ via

$$T(t)\mu(\cdot) \triangleq \int p(\xi, \cdot)\mu(d\xi)$$

where p is as in (b) above. It is then the case that $\mathcal{M}$ is metrizable as a complete metric space and $(\mathcal{M}, T)$ is a semidynamical system (see [63, Saperstone]). An important idea in semidynamical systems is that of an invariant point (see Appendix C for definition). In the case of $(\mathcal{M}, T)$, μ is an invariant point if $T(t)\mu(\cdot) = \mu(\cdot) \ \forall t \geq 0$ so that if $\mathbf{x}(0) \sim \mu_0$ and μ_0 is an invariant point, then $\mathbf{x}(t) \sim \mu_0$, $\forall t \geq 1$, i.e. $\{\mathbf{x}(t)\}$ is a strictly stationary Markov process. Such measures are important in describing the asymptotic behaviour of Markov processes, and in view of their defining property, they are called *invariant measures*.

Having painted the picture of the evolution of the process $\{\mathbf{x}(t)\}$ and its distributions $\{\mu_t\}$, consider the following key stability result for the controlled system (4.42).

Proposition 4.1 [58, Meyn and Caines]

For the system (4.42) with $\mathbf{x}(0) \sim \delta_0$, the unit probability mass at zero,

$$\limsup_{t \to \infty} E[\|\mathbf{x}(t)\|^2] < \infty \text{ iff } \sigma_\epsilon^2 < 1.$$

∎

In particular, one can establish that with $\mathbf{x}(0) \sim \delta_0$

$$\sigma_\epsilon^2 < V(t|t-1) < \frac{\sigma_\epsilon^2}{1 - \alpha^2} \text{ a.s.} \quad \forall t \geq 2, \tag{4.43}$$

a result that will be used below.

Definition 4.1

A Borel measurable function $f(\cdot)$ on $\Re^n$ is a moment if $f \geq 0$ and

$$\lim_{n \to \infty} \left(\inf_{\mathbf{x} \in \Re^n \backslash K_n} f(\mathbf{x}) \right) = \infty$$

for some sequence of compact sets $K_n \uparrow \Re^n$. ∎

Theorem 4.3 [10, Beneš]

Suppose the stochastic transition function p of a process $\{\mathbf{x}(t)\}$ satisfies the Feller property. The following are then equivalent.

(i)   There exists an invariant measure for $\{\mathbf{x}(t)\}$.

(ii) There exists a moment f such that for some initial distribution μ_0 of $\mathbf{x}(0)$

$$\sup_{t \geq 0} \mathrm{E}[f(\mathbf{x}(t))] < \infty \qquad \text{or} \qquad \sup_{N \geq 1} \frac{1}{N} \sum_{t=1}^{N} \mathrm{E}[f(\mathbf{x}(t))] < \infty.$$

(iii) For some initial distribution μ_0 of $\mathbf{x}(0)$ and $\forall \epsilon > 0 \; \exists K_\epsilon$ compact s.t.

$$\inf_{t \geq 0} U(t) \mu_0(K_\epsilon) > 1 - \epsilon.$$

∎

Firstly, note that theorem 4.3 establishes the existence of an invariant measure for the example under study. Part (iii) of the theorem is actually the definition of a *tight* sequence of measures and can be established by using Chebyshev's inequality and the boundedness of $\mathrm{E}[\|\mathbf{x}(t)\|_i^2]$. Further, note that $\mathcal{F}_\nu^I$ and $\mathcal{F}_\nu^D$ of (c) above (with $\Re^3$ and $\mathcal{B}^3$ replaced by $\Re^n$ and $\mathcal{B}^n$ respectively) are σ-algebras and divide the state space of $\{\mathbf{x}(t)\}$ in two families of sets. If $\mathbf{x}(0) \, \epsilon \, A \, \epsilon \, \mathcal{F}_\nu^I$ then $\mathbf{x}(t) \, \epsilon \, A$, $\forall t \geq 1$ w.p. 1 (μ_∞), while if $\mathbf{x}(0) \, \epsilon \, \mathcal{B}^n \backslash A$ then $\mathbf{x}(t) \, \epsilon \, \mathcal{B}^n \backslash A$, $\forall t \geq 1$. Members of $\mathcal{F}_\nu^I$ are called *invariant sets* for this reason. Looking at $\mathcal{F}_\nu^D$, its members are sets A corresponding to which their exist sets $B \, \epsilon \, \mathcal{B}^n$ such that if $\mathbf{x}(t) \, \epsilon \, B$ then $\mathbf{x}(t+1) \, \epsilon \, A$ w.p. $1(\mu_\infty), \forall t \geq 1$. Such sets are called *deterministic sets* since once entered, the next state is determined a.s. Trivially, $\mathcal{F}_\nu^I \subset \mathcal{F}_\nu^D$ and if no *proper* deterministic subsets of $\mathcal{B}^n$ exist, $\mathcal{F}_\nu^I = \mathcal{F}_\nu^D = \{\mathcal{B}^n, \emptyset\}$. In light of (c), the following theorem establishes the uniqueness of an invariant measure μ_∞ for the example under study.

79

Theorem 4.4 [38, Foguel]

If $p(\xi, A)$ is continuous for all $A \in B^n$ and there are no proper invariant subsets of B^n, then there is at most one invariant measure on B^n. ∎

The asymptotic distribution of $\{x(t)\}$ is specified by the following.

Theorem 4.5 [38, Foguel]

Suppose there are no proper deterministic subsets of B^n. Then for all $A \in B^n$ and any initial distribution $\mu_0 \prec \mu_\infty$

$$\lim_{t \to \infty} P(x(t) \in A) = \mu_\infty(A)$$

∎

With mild assumptions on $\{x(t)\}$, the result above also holds if μ_0 is replaced by δ_ξ for a.s. (μ_∞) $\xi \in \Re^n$.

Asymptotic sample mean results for a large class of functions of the state can finally be established using the following two theorems which assume the existence of an invariant measure μ_∞.

Theorem 4.6 [34, Doob]

Let $f \in L^1(\Re^n, B^n, \mu_\infty)$. Then

$$\lim_{N \to \infty} \frac{1}{N} \sum_{t=1}^{N} f(x(t)) = E_\infty[f(x(0)|\mathcal{F}_{\mu_\infty}^I] \text{ a.s. for a.a.}(\mu_\infty) \ x(0) \in \Re^n$$

or a.s. when $x(0) \sim \mu_0$ with $\mu_0 \prec \mu_\infty$. ∎

Theorem 4.7 [48, Hopf]

Let $f \in L^1(\Re^n, B^n, \mu_\infty)$. Then

$$\lim_{N \to \infty} \frac{1}{N} \sum_{t=1}^{N} E[f(x(t))|x(0) = \xi] = E_\infty[f(x(0))|\mathcal{F}_{\mu_\infty}^I] \text{ a.s. for a.a.}(\mu_\infty) \ \xi \in \Re^n.$$

∎

To see how these results apply to the example under study suppose that $x(0) \sim \delta_0$ (whence (4.39) generates conditional expectations) and denote the unique invariant measure by μ_∞. Taking conditional expectations in (4.40) gives

$$E[y^2(t)|\mathcal{F}_{t-1}^y] = V(t|t-1)y^2(t-1) + \sigma_w^2 \tag{4.44}$$

where the assumptions (i) and (ii) were used. Multiplying (4.44) by (4.39b) and using the $\mathcal{F}^y_{t-1}$ measurability of $V(t+1|t)$ gives

$$E[V(t+1|t)y^2(t)|\mathcal{F}^y_{t-1}] = \alpha^2\sigma_w^2 V(t|t-1) + \sigma_e^2 E[y^2(t)|\mathcal{F}^y_{t-1}].$$

Taking expected values above results in

$$E[V(t+1|t)y^2(t)] = \alpha^2\sigma_w^2 E[V(t|t-1)] + \sigma_e^2 E[y^2(t)] \qquad (4.45)$$

while doing the same on (4.44) yields

$$E[V(t+1|t)y^2(t)] = E[y^2(t+1)] - \sigma_w^2. \qquad (4.46)$$

Taking Cesaro sums of (4.45) and (4.46), passing to the limit, and using theorem 4.7 gives

$$(1-\sigma_e^2)E_\infty[y^2(0)] = \sigma_w^2 + \alpha^2\sigma_w^2 E_\infty[V(1|0)],$$

and using (4.43) one finally obtains

$$\frac{\sigma_w^2}{1-\sigma_e^2}(1+\alpha^2\sigma_e^2) \le E_\infty[y^2(0)] \le \frac{\sigma_w^2}{1-\sigma_e^2}\left(1+\frac{\alpha^2\sigma_e^2}{1-\alpha^2}\right). \qquad (4.47)$$

This inequality is important in view of the fact that theorem 4.6 yields the limit

$$\lim_{N\to\infty}\frac{1}{N}\sum_{t=1}^{N}y^2(t) = E_\infty[y^2(0)] \qquad \text{a.s.} \qquad (4.48)$$

whenever $\mu_0 = \delta_\xi$ where $\xi \in \mathcal{R}^3\backslash M$ and $\mu_\infty(M) = 0$ or when $\mu_0 \prec \mu_\infty$. Moreover, for each $\varepsilon > 0$, theorems 4.5 and 4.6 give

$$\lim_{t\to\infty} P(|y(t)| > \varepsilon) = \mu_\infty(y^2(0) > \varepsilon^2)$$

$$(4.49)$$

$$\le \frac{1}{\varepsilon^2}E_\infty[y^2(0)].$$

For this example, [58, Meyn and Caines] shows that results (4.48) and (4.49) hold for all initial distributions μ_0 irrespective of whether equations (4.39) generate conditional expectations or not. Note that setting σ_e^2 to zero (i.e. the constant parameter case) results in

$$\lim_{N\to\infty}\frac{1}{N}\sum_{t=1}^{N}y^2(t) = \sigma_w^2 \qquad \text{a.s.} \qquad (4.50)$$

which is the familiar result obtained in Chapter 2. Theorem 4.7 yields the asymptotic value of a criterion similar to the one used in [41, Goodwin et.al], namely,

$$\lim_{N \to \infty} \frac{1}{N} \sum_{t=1}^{N} E[y^2(t)|x(0) = \xi] = E_\infty[y^2(0)]$$

which also reduces to σ_w^2 in the constant parameter case. It must be noted however that, for technical reasons, the result (2.39c) cannot be obtained by setting $\sigma_\epsilon^2 = 0$.

Chapter 5   Computer Simulations of Adaptive
                                          Control Algorithms

5.1 Introduction

This chapter presents simulations of some of the algorithms presented in Chapter 4. Specifically, the first few examples examine the robustness of the MLS algorithm when the parameters of the system model are subjected to bounded random variations about their mean values. Performance deterioration is demonstrated when these variations become large enough, and the results of theorem 4.2 are shown to be conservative. The second set of examples show simulations of the control of the system (4.33) using (4.36) and (4.39). Via Proposition 4.1, results expressed in (4.47) and (4.48) are verified and instability is demonstrated when the condition in the aforementioned proposition is violated.

5.2 Simulation Descriptions

The introductory comments of Chapter 3 regarding pseudo-random number generators and algorithm implementations apply equally to this chapter without further comment. Each simulation consists of 1000 sample points.

Example 5.1

Consider the MLS (Modification 1) self-tuning control algorithm presented in theorem 4.2 for the ARX system

$$A(z) = 1 + a_1(t)z, \quad a_1(t) = a_1 + (\Delta a_1)(t), \ a_1 = 2.0,$$
$$B(z) = b_0 + b_1 z, \quad b_0 = 1.0, \ b_1 = 0.5.$$

The constants K_1 and K_2 in Modification 1 of the MLS algorithm (2.31) are chosen as 10.0 and 0.9 respectively so that via (4.9d) and (4.11b), $|(\Delta a_1)(t)| < \alpha < \frac{1}{21}$, represents the allowable system parameter variation for which the results of theorem 4.2 hold. In this first example, a disturbance uniformly distributed in $(-0.1, 0.1)$ is added to a_t so that $|(\Delta a_1)(t)| < 0.1$. The disturbance sequence $\{w(t)\}$ is taken as $N(0, 0.5)$ while the delay is set at $d = 1$. The initial parameter estimate is arbitrarily set at

$$\widehat{\vartheta}(0) = [-1.0, 0.5 \ 0.1]^T,$$

with $x_0 = 0$ and $y^*(t) \equiv 0$.

Figures 5.1 (a) - (j) pertain to the example just described. Figures (a) and (b) show the adaptive and non-adaptive output sample paths respectively, the responses being quite similar. Stabilization and tracking are evident with the average squared tracking error (ASTE) shown in figures (c) and (d). While the adaptive controller displays a transient causing a large initial tracking error, the latter converges to approximately the same value as the non-adaptive controller, attesting to the robustness of the self-tuning mechanism, even when the parameter variations are approximately twice the maximum specified in theorem 4.2. Figures (e), (f), and (g) show the estimated parameters $\widehat{\vartheta}_1(t), \widehat{\vartheta}_2(t)$, and $\widehat{\vartheta}_3(t)$ while figure (h) displays a plot of $\widetilde{\vartheta}_1(t) = a_1(t) - \widehat{\vartheta}_1(t)$. Figures (i) and (j) show the adaptive and non-adaptive output sample path autocorrelations. These are almost identical. ∎

Example 5.2

Continuing the above example, the parameter variation $\Delta a_1(t)$ is now increased to $|(\Delta a_1)(t)| \leq 1.0$, this being the only change. Figure 5.2(a) shows the resulting output sample path and figure 5.2(b) the ASTE. The system has again been stabilized although a larger transient behaviour is observed. Clearly, as was already evident in the previous example, the bound α specified in (4.11c) is very tight, while that of (4.11b) too loose. The quantities $\widehat{\vartheta}_1, \widehat{\vartheta}_2, \widehat{\vartheta}_3, \widetilde{\vartheta}_1(t) = a_1(t) - \widehat{\vartheta}_1(t)$, and the output sample path autocorrelation are shown in figures 5.2 (c) - (g) respectively. Wherever they appear. dashed lines are plots related to the non-adaptive controller. These of course do not change from example to example as the bound on $(\Delta a_1)(t)$ increases. ∎

Example 5.3

Further increasing $\Delta a_1(t)$ to $|(\Delta a_1)(t)| \leq 2.0$, 100 times the amount specified by theorem 4.2, results in an increasingly poor performance. Plots analogous to those in figures 5.2 (b) - (g) are shown in figures 5.3 (c) - (h). Figure 5.3 (a) shows the output sample path $\{y(t)\}$ while 5.3 (b) displays $\ell n(y(t))$ for positive $y(t)$ and $-\ell n(-y(t))$ for negative $y(t)$, values of $y(t)$ in $[-1,1]$ remaining unchanged. The output variations about the zero reference point are clearly quite unacceptable in view of the 0.5 variance of the disturbance sequence. It is noteworthy however that the ASTE decreases steadily from its large initial value thereby showing the continued but unacceptably slow effort of the controller to stabilize the system. ∎

Example 5.4

As a final demonstration of how the performance of the controller deteriorates, the parameter variation in this continuation of example 5.1 is taken as $|(\Delta a_1)(t)| \leq 2.5$. Figures 5.4 (a) - (h) are completely analogous to those of figure 5.3. An increase of parameter variation to 2.75 results in numbers beyond the range of the computer. ∎

Example 5.5

In this example variations of the moving average parameters are also allowed. Specifically, with reference to example 5.1, the $B(z)$ polynomial now becomes

$$B(z) = b_0(t) + b_1(t)z, \quad b_0(t) = b_0 + (\Delta b_0)(t), \quad b_0 = 1.0,$$
$$b_1(t) = b_1 + (\Delta b_1)(t), \quad b_1 = 0.5.$$

The range of variations in each of the parameters is 0.1 i.e. each of $\Delta a_1(t), \Delta b_0(t),$ and $\Delta b_1(t)$ are uniformly distributed in (-0.1, 0.1). Plots of the output sample path and the ASTE are shown in figure 5.5(a) and (b) respectively. Stabilization and tracking are observed with an initial large transient resulting in a large tracking error which dies out quite slowly. Figures 5.5 (c) - (i) show plots of $\widehat{\vartheta}_1(t), \widetilde{\vartheta}_1(t) = a_1(t) - \widehat{\vartheta}_1(t), \widehat{\vartheta}_2(t), \widetilde{\vartheta}_2(t) =$

$b_0(t) - \hat{\vartheta}_2(t), \hat{\vartheta}_3(t), \tilde{\vartheta}_3(t) = b_1(t) - \tilde{\vartheta}_3(t)$, and the output sample path autocorrelation respectively. ∎

In constrast with the previous examples, the following ones show the performance of a controller which is designed for time varying systems.

Example 5.6

Consider the system described by

$$y(t) + \vartheta(t)y(t-1) = bu(t-1) + w(t), \quad t \geq 1$$

with initial condition $x_0 = 0$ and $w(t) \sim N(0, \sigma_w^2)$ $\forall t \geq 1$. Further let $\vartheta(t)$ be given by

$$\vartheta(t) = \alpha\vartheta(t-1) + v(t), \quad t \geq 1,$$

with $\vartheta(0) = 0$ and $v(t) \sim N(0, \sigma_v^2)$. As described in section 4.5, the control strategy (4.36) together with (4.38) and (4.39) is adopted. The coefficient b of $u(t-1)$ is assumed equal to 1. Further, $\vartheta^\circ(1|0) = 10.0, V(1|0) = 1.0, \{y^*(t)\} \equiv 0, \sigma_w^2 = 1.4$, and $\sigma_v^2 = 0.6$. Figure 5.6 shows the simulation results with dashed line curves representing the non-adaptive controller as always. Figures (a) - (d) are sample path plots while (e) - (g) display empirical un-normalized densities of $y(300), y(600)$, and $y(900)$ obtained from data generated by simulating 500 sample paths of 1000 samples each. Table 5.1 below shows the results of the Kolmogorov one-sample test used to assess the normality of $y(300), y(600)$, and $y(900)$ when compared to theoretical normal distributions determined by empirical means and variances. The latter are also recorded in Table 5.1.

	$\sqrt{n}D_n$	α	μ	σ^2
$w(300)$	0.70260	0.70690	-0.0101	1.3676
$y(300)$	2.2893	0.000056	0.0493	5.7912
$w(600)$	0.6557	0.7829	0.0642	1.3713
$y(600)$	1.3506	0.0521	0.0826	4.1571
$w(900)$	0.5636	0.9085	0.000081	1.3991
$y(900)$	0.8508	0.4641	0.1183	3.4363

Table 5.1 Statistical results for Example 5.6

The output sample path in figure (a) shows tracking about the reference trajectory while figure (b) shows the adaptive and non-adaptive ASTE's. The latter are in agreement with theoretical predictions; equation (4.50) relates to the non-adaptive ASTE while (4.47) and (4.48) predict that $4.025 \leq$ Adaptive ASTE ≤ 4.2. The last values plotted in Figure (b) are 1.452 for the non-adaptive case and 4.063 for the adaptive case. Figures (c) - (d) show $\tilde{\vartheta}_1(t|t-1)$ and the output sample path autocorrelation $\widehat{R}_y(t)$. Note that $\widehat{R}_y(0) = 4.0727$, quite close to the ASTE mentioned above. ∎

Example 5.7

Here the only change from the previous example is that $\alpha = 0.99$. This enables the parameter $\vartheta(t)$ to vary more gradually so that it stays longer at a certain level, possibly one that gives an unstable system (i.e. $|\vartheta(t)| \geq 1$). The simulations found in figure 5.7 (a) - (g) are in one to one analogy with those of figure 5.6. Here, (4.47) and (4.48) predict that $5.56 \leq$ Adaptive ASTE ≤ 106.93. The last computed value of the adaptive ASTE is 13.85 while the value of the output sample path autocorrelation $\widehat{R}_y(0) = 9.0558$. Table 5.2 below shows results on cross-sectional data obtained from 500 sample path simulations.

	$\sqrt{n}D_n$	α	μ	σ^2
$w(300)$	0.70260	0.70690	−0.0101	1.3676
$y(300)$	3.4460	0.0	0.1498	14.130
$w(600)$	0.6557	0.7829	0.0642	1.3713
$y(600)$	1.5020	0.0219	0.0825	6.4804
$w(900)$	0.5636	0.9085	0.000081	1.3991
$y(900)$	1.8212	0.0026	0.0779	7.1819

Table 5.2 Statistical results for Example 5.7

Special attention should be given to figure 5.7 (a) which demonstrates the highly non-linear behaviour of the controlled system. The pronounced spikes indicate that it is highly unlikely that this output sample path could have been generated by a linear stochastic system. The plausible interpretation is that, after each burst, the 'learning' ability of the controller stabilizes the system until substantial changes in the parameters of the plant require updating the controller for satisfactory performance. This cycle then repeats itself. ■

The following two examples demonstrate situations where σ_v^2 is greater than 1, i.e. situations which are theoretically non-stabilizable.

Example 5.8

Starting with example 5.6, change σ_v^2 to 2.5. The results are shown in figures 5.8 (a) - (g). Figure (a) shows the output $y(t)$ while (b) shows $\ell n(y(t))$ for positive $y(t)$ and $-\ell n(-y(t))$ for negative $y(t)$, values in $[-1,1]$ remaining unaffected. Figures (c) and (d) show plots of the ASTE and its associated logarithmic plot. Finally figures (e) and (f) show $\tilde{\vartheta}_1(t|t-1)$ and the output sample path autocorrelation $\hat{R}_y(\tau)$. While the output contains unacceptably large bursts, its behaviour cannot be said to be unstable. Rather, the non-linear behaviour mentioned in the previous example seems to be even more prevalent and the system seems to be stabilizable. ■

Example 5.9

The parameter noise variance is increased to $\sigma_v^2 = 3.5$ and the results are shown in Figure 5.9. Figure 5.9(a) shows the explosively unstable non-linear output behaviour where the non-linear ordinate scale should be noted. Figure 5.9(b) shows the exploding ASTE, while Figures 5.9(c) and (d) show $\tilde{\vartheta}_1(t|t-1)$ and the output sample path autocorrelation $\hat{R}_y(\tau)$. ■

5.3 Simulation Results

The following pages display the plots corresponding to the examples described above.

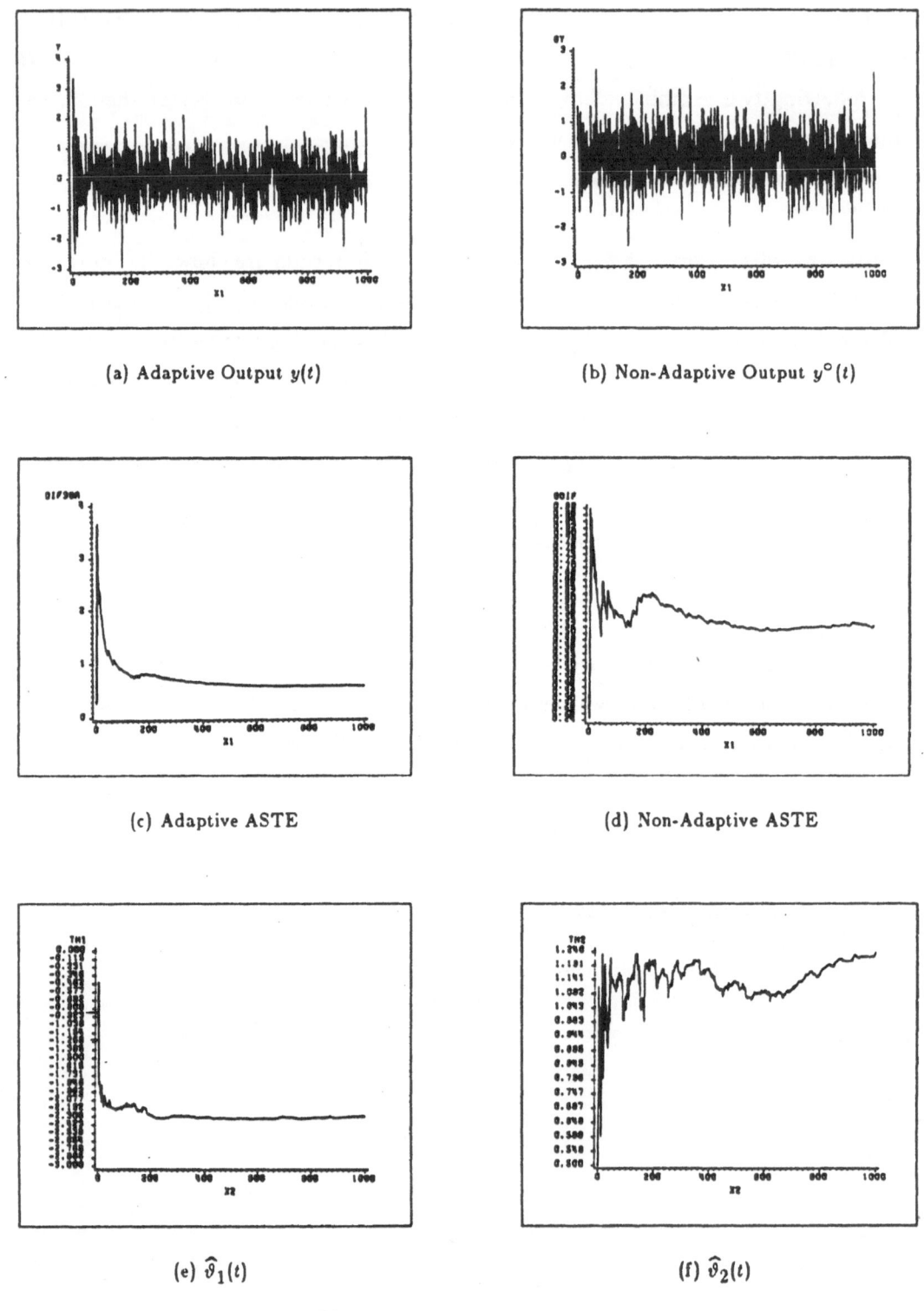

(a) Adaptive Output $y(t)$

(b) Non-Adaptive Output $y^\circ(t)$

(c) Adaptive ASTE

(d) Non-Adaptive ASTE

(e) $\widehat{\vartheta}_1(t)$

(f) $\widehat{\vartheta}_2(t)$

Figure 5.1 Example 5.1 (MLS)

89

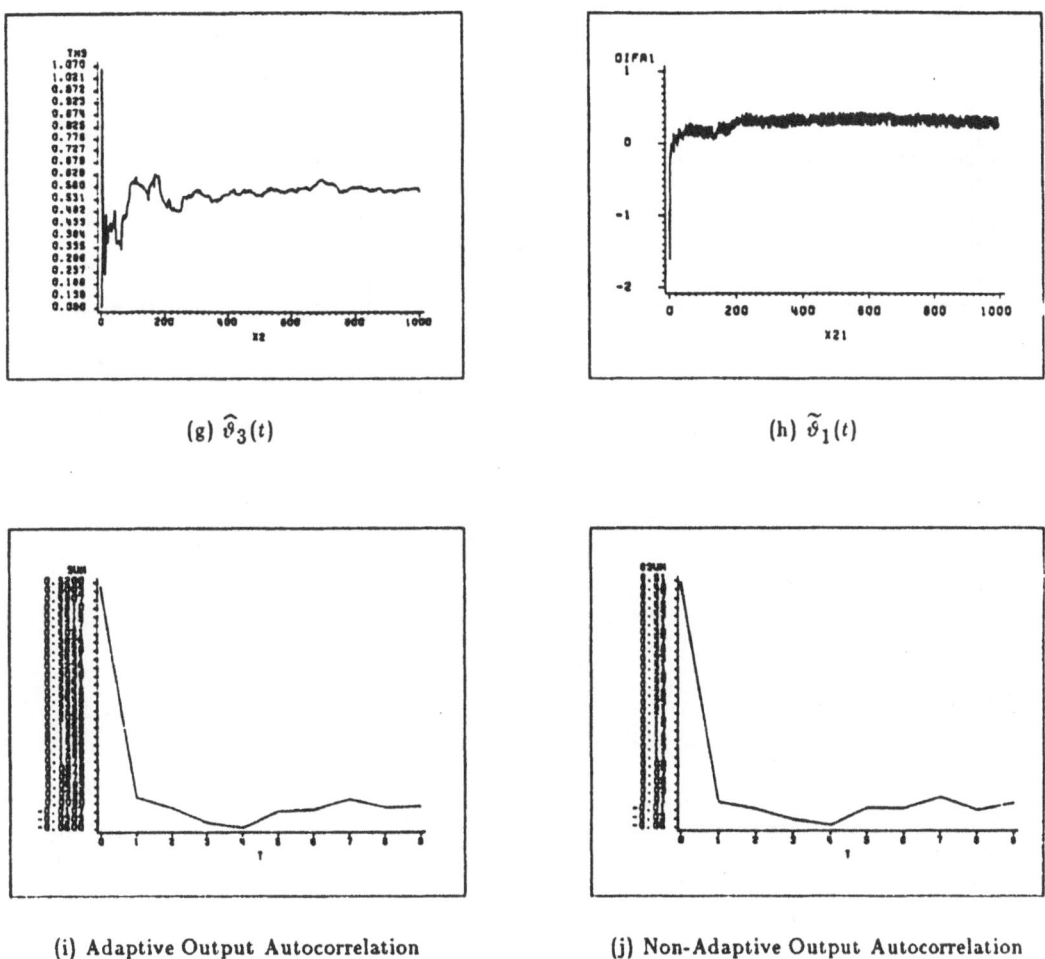

(g) $\widehat{\vartheta}_3(t)$

(h) $\widetilde{\vartheta}_1(t)$

(i) Adaptive Output Autocorrelation

(j) Non-Adaptive Output Autocorrelation

Figure 5.1 Example 5.1 (MLS)

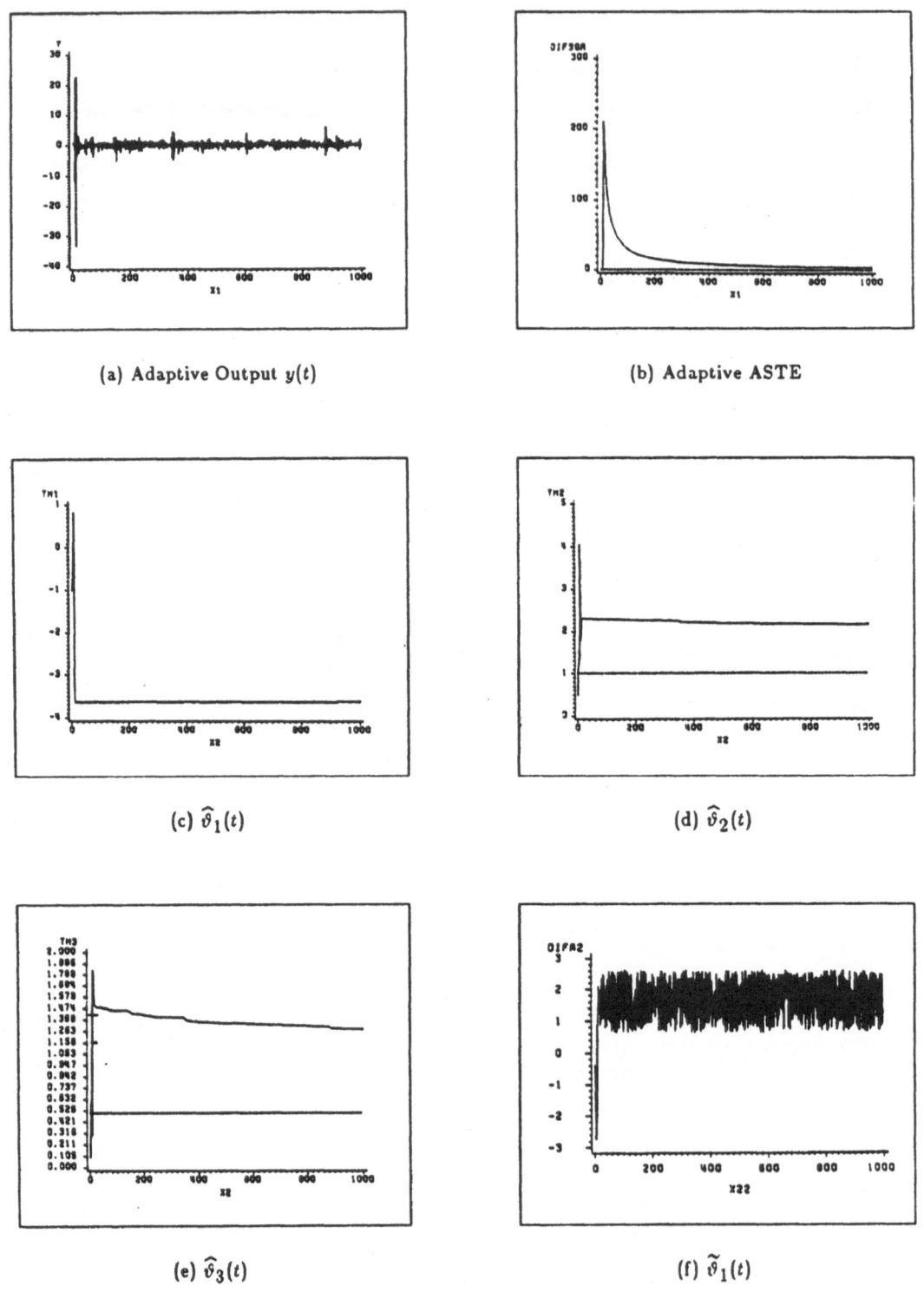

(a) Adaptive Output $y(t)$

(b) Adaptive ASTE

(c) $\widehat{\vartheta}_1(t)$

(d) $\widehat{\vartheta}_2(t)$

(e) $\widehat{\vartheta}_3(t)$

(f) $\widetilde{\vartheta}_1(t)$

Figure 5.2 Example 5.2 (MLS)

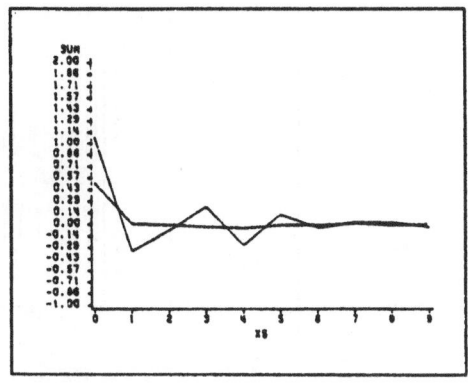

(g) Adaptive Output Autocorrelation

Figure 5.2   Example 5.2 (MLS)

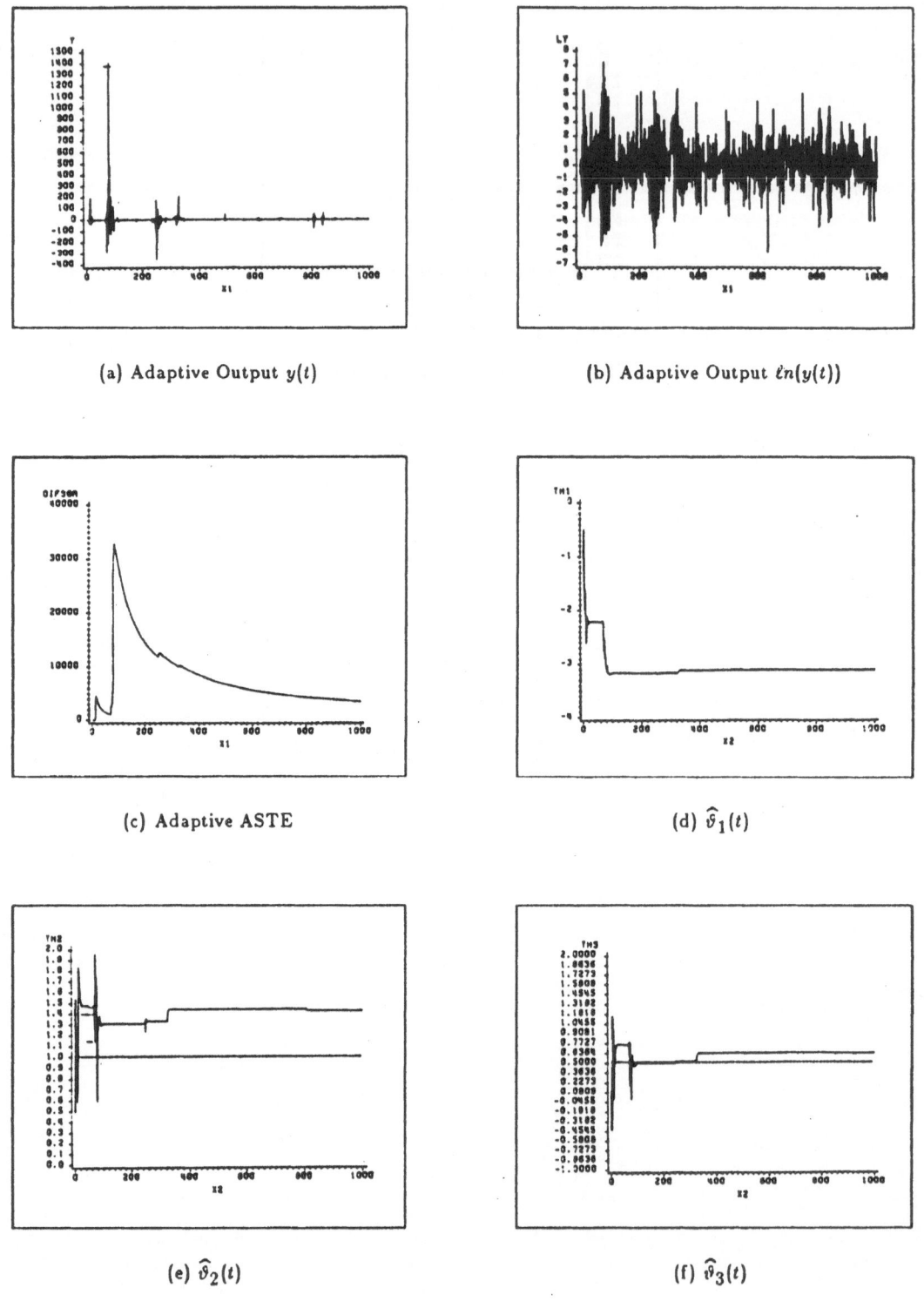

(a) Adaptive Output $y(t)$

(b) Adaptive Output $\ell n(y(t))$

(c) Adaptive ASTE

(d) $\widehat{\vartheta}_1(t)$

(e) $\widehat{\vartheta}_2(t)$

(f) $\widehat{\vartheta}_3(t)$

Figure 5.3 Example 5.3 (MLS)

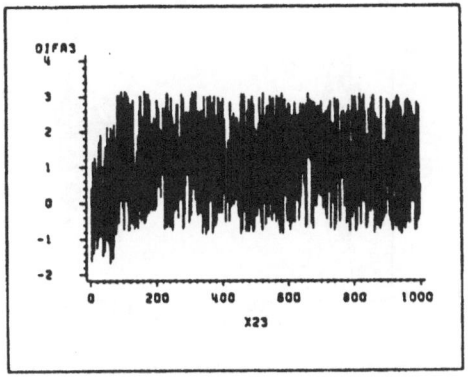

(g) $\widetilde{\vartheta}_1(t)$

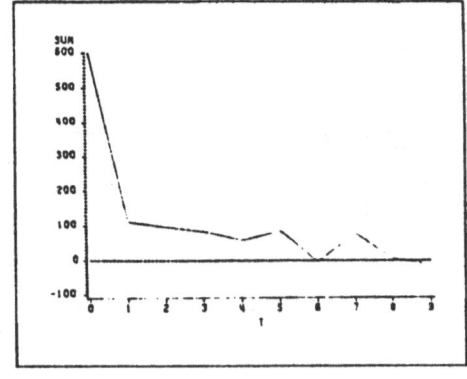

(h) Adaptive Output Autocorrelation

Figure 5.3   Example 5.3 (MLS)

94

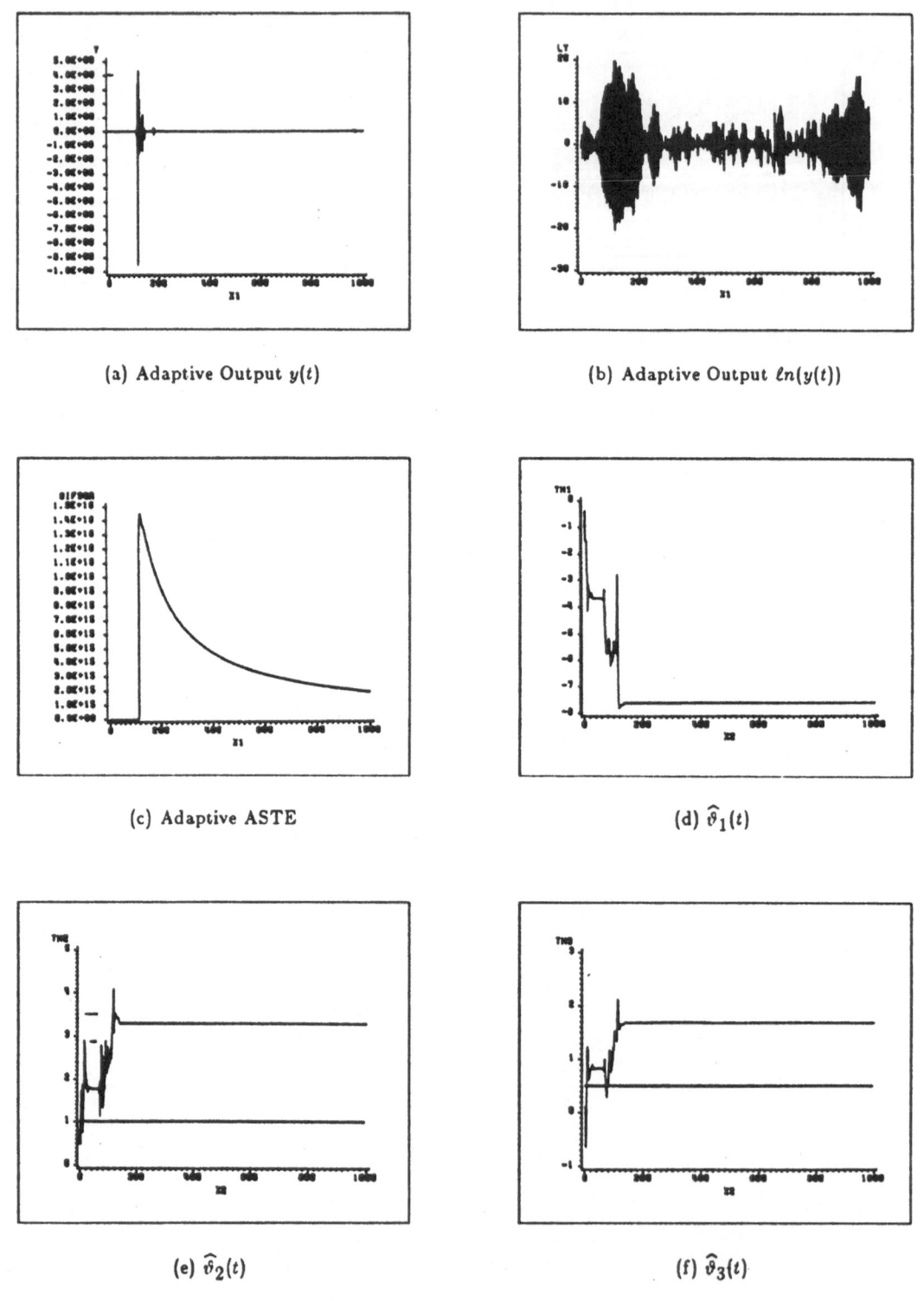

(a) Adaptive Output $y(t)$

(b) Adaptive Output $ln(y(t))$

(c) Adaptive ASTE

(d) $\widehat{\vartheta}_1(t)$

(e) $\widehat{\vartheta}_2(t)$

(f) $\widehat{\vartheta}_3(t)$

Figure 5.4 Example 5.4 (MLS)

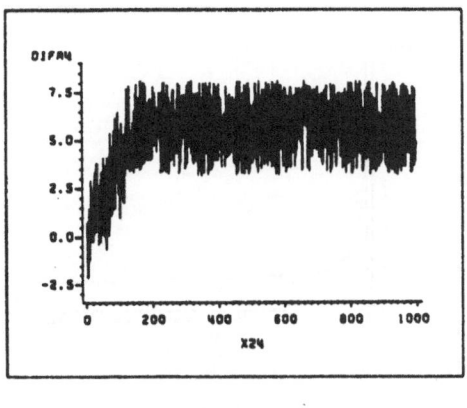

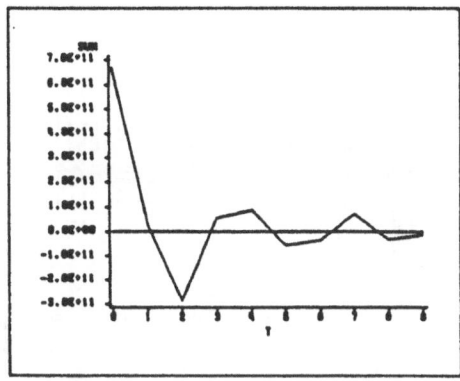

(g) $\widetilde{\vartheta}_1(t)$

(h) Adaptive Output Autocorrelation

Figure 5.4   Example 5.4 (MLS)

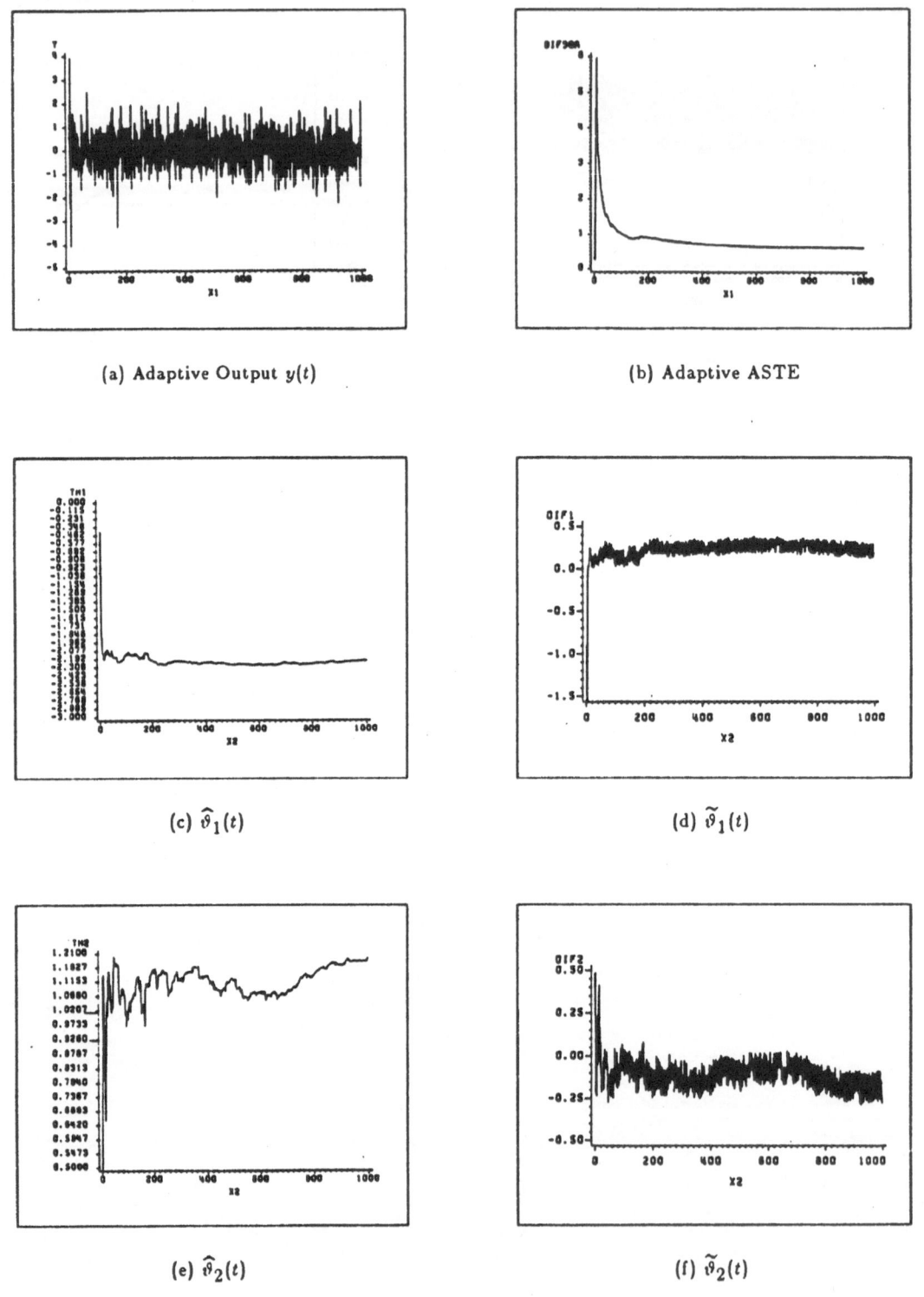

(a) Adaptive Output $y(t)$

(b) Adaptive ASTE

(c) $\widehat{\vartheta}_1(t)$

(d) $\widetilde{\vartheta}_1(t)$

(e) $\widehat{\vartheta}_2(t)$

(f) $\widetilde{\vartheta}_2(t)$

Figure 5.5 Example 5.5 (MLS)

97

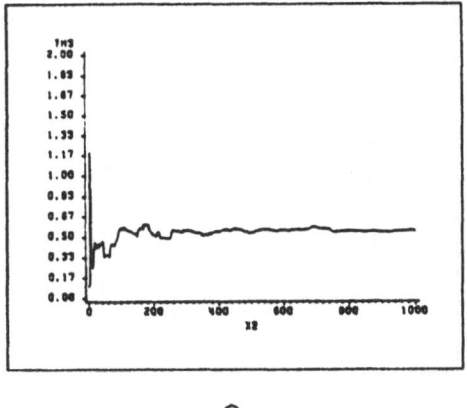

(g) $\widehat{\vartheta}_3(t)$

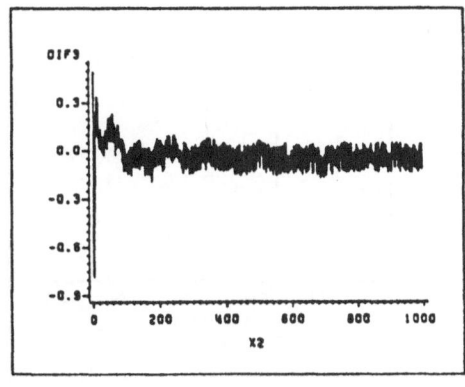

(h) $\widetilde{\vartheta}_3(t)$

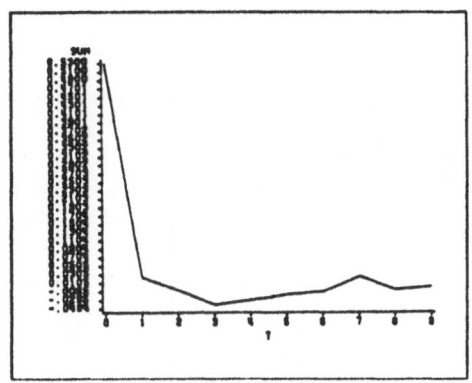

(i) Adaptive Output Autocorrelation

Figure 5.5 Example 5.5 (MLS)

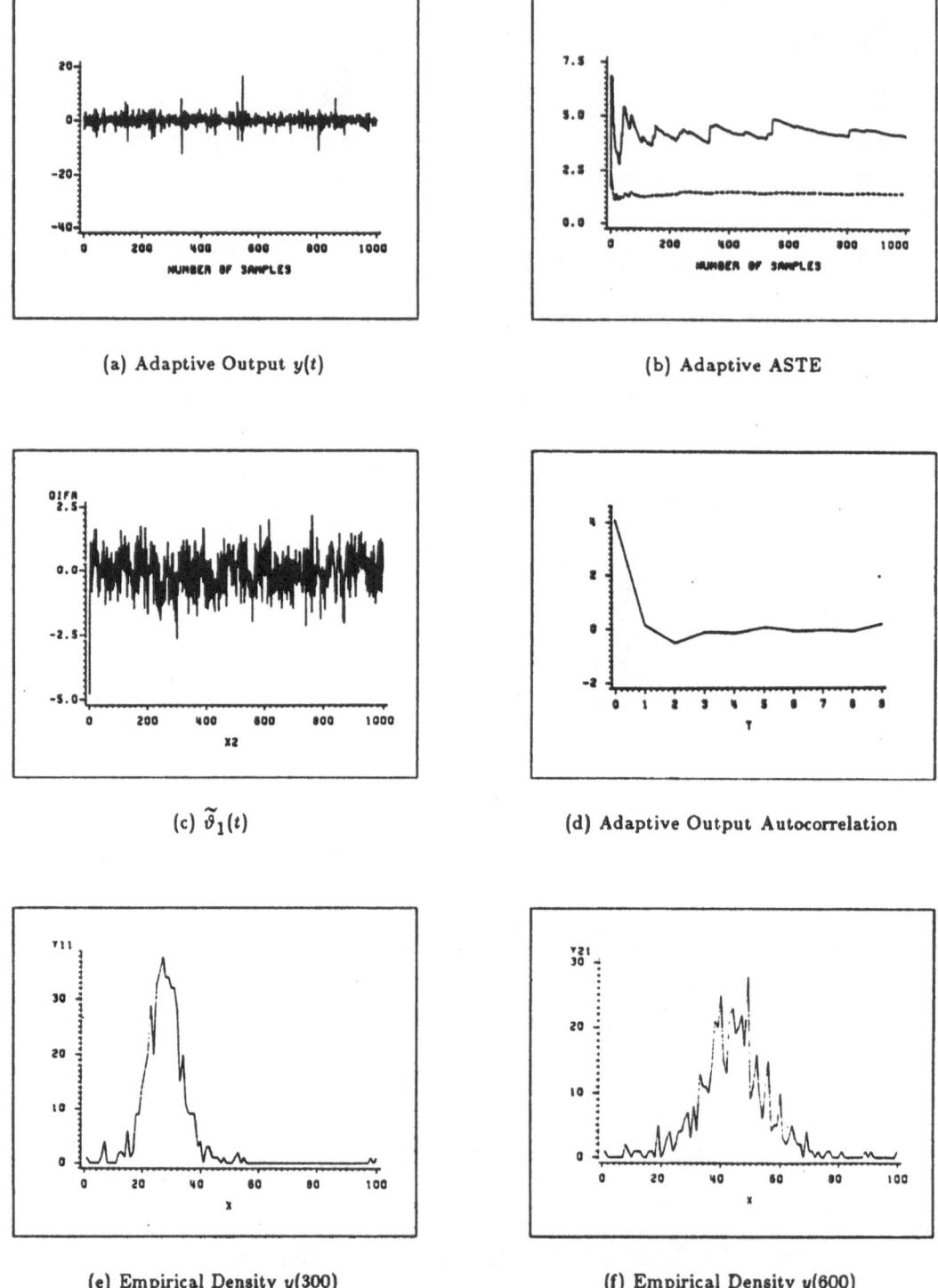

(a) Adaptive Output $y(t)$

(b) Adaptive ASTE

(c) $\widetilde{\vartheta}_1(t)$

(d) Adaptive Output Autocorrelation

(e) Empirical Density $y(300)$

(f) Empirical Density $y(600)$

Figure 5.6 Example 5.6 (Kalman Filter)

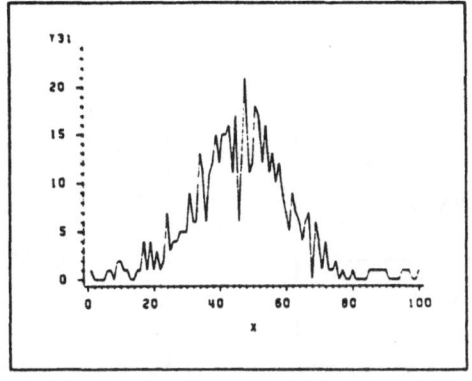

(g) Empirical Density $y(900)$

Figure 5.6   Example 5.6 (Kalman Filter)

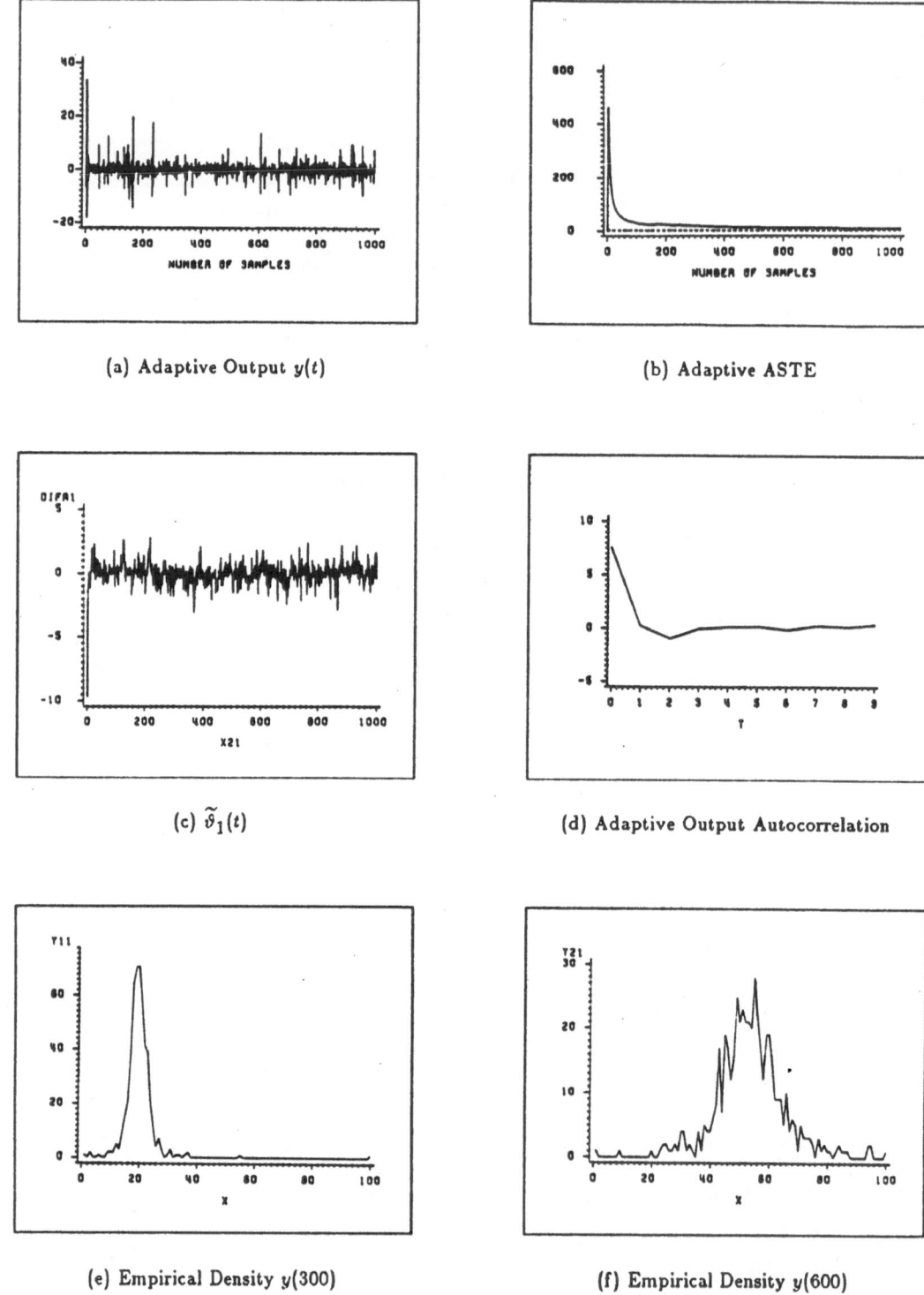

(a) Adaptive Output $y(t)$

(b) Adaptive ASTE

(c) $\widetilde{\vartheta}_1(t)$

(d) Adaptive Output Autocorrelation

(e) Empirical Density $y(300)$

(f) Empirical Density $y(600)$

Figure 5.7   Example 5.7 (Kalman Filter)

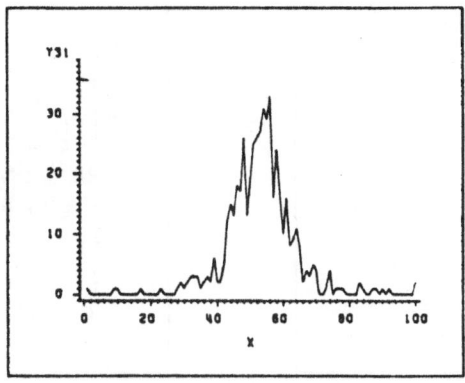

(g) Empirical Density $y(900)$

Figure 5.7 Example 5.7 (Kalman Filter)

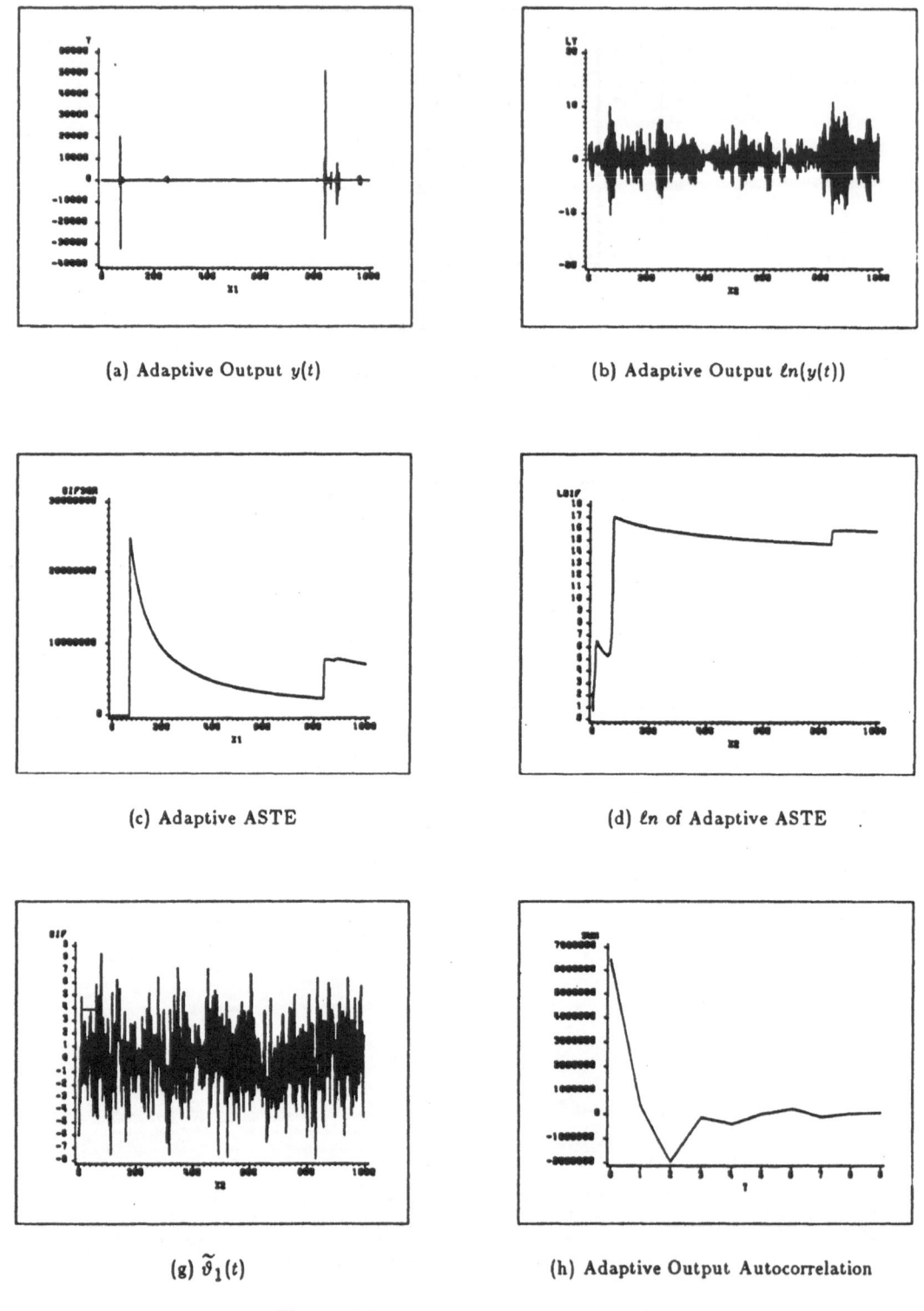

(a) Adaptive Output $y(t)$

(b) Adaptive Output $\ell n(y(t))$

(c) Adaptive ASTE

(d) ℓn of Adaptive ASTE

(g) $\widetilde{\vartheta}_1(t)$

(h) Adaptive Output Autocorrelation

Figure 5.8   Example 5.8 (Kalman Filter)

103

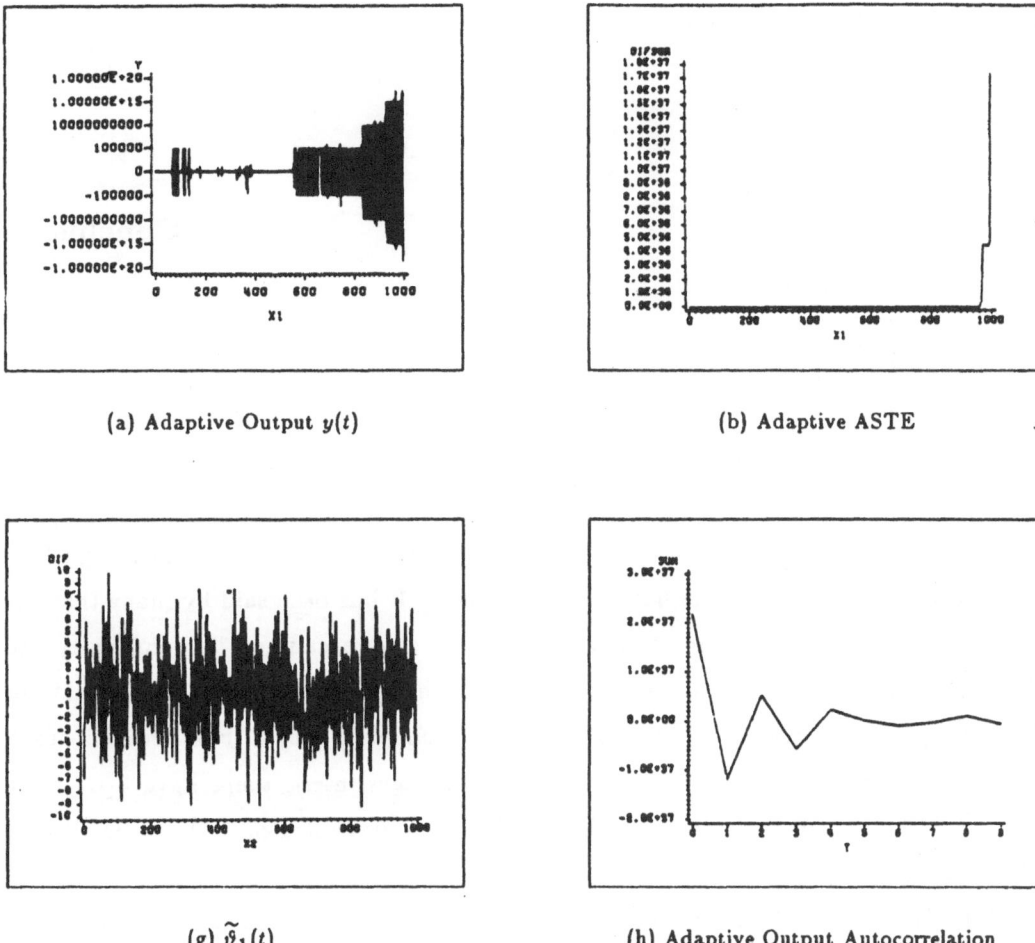

(a) Adaptive Output $y(t)$

(b) Adaptive ASTE

(g) $\widetilde{\vartheta}_1(t)$

(h) Adaptive Output Autocorrelation

Figure 5.9 Example 5.9 (Kalman Filter)

Chapter 6   Conclusion

6.1   Overview

A daptive control has a history which spans almost a whole generation during which a vast number of papers have been published. It has been said by many that the developments in this field have been comparatively slow because of its complexity and difficulty (e.g. [59, Narendra and Monopoli]). Although many successful applications have been reported (see e.g. [59, Narendra and Monopoli], [40, Gilbart and Winston], [22, Cegrell and Hedqvist], [35, Dumont and Bélanger] etc.), there have been some failures ([60, NASA]) and these may have resulted in psychological setbacks in the use of adaptive control.

Since its birth, adaptive control theorists have been trying to answer two fundamental questions. The first question is 'what is adaptive control?', some answers to which have been discussed in the introduction. To put things in perspective, consider the following. In the most general case, control problems are posed for non-linear plants subject to random disturbances and it is at this level of generality that one finds adaptive control problems. However, for optimal control, the intractability of the resulting Bellman equation forces one to focus on the class of LQG problems, a class which includes all the linear problems of classical feedback control theory [47, Harris and Billings]. The design of an optimal controller for the LQG problem must still deal with the fact that the coefficients of the assumed linear model are unknown. This leads to the separation of the problem into subproblems of identification and control, or to its conversion to a search of an optimal controller for a non-linear system where, however, the nonlinearities are of a tractable nature. The former approach — under which this study may be classified — can be viewed in two ways. The first is the complete separation of identification and control, i.e. identify the plant and then design a controller. This route does not involve adaptation and is not always possible — consider the case of an unstable unknown plant. The second route is the adaptive one, namely, to estimate the uncertain coefficients of the assumed model and use the resulting information to continuously update the controller. It is within this approach that one finds self-tuning, model-reference, and dual controllers. In the way described above, the answer to the first fundamental question is that adaptive control is *an approach to designing suboptimal controllers for non-linear stochastic problems*, the key feature of this approach being the creation of algorithms which recursively estimate the parameters — or more generally the information — necessary for a satisfactory solution of the control problem.

The second fundamental question, namely, 'when is adaptive control useful?' is perhaps easier to answer since only practical experience with real life problems can ultimately provide an answer. A point to consider however, and one brought up by many authors, is whether the plant variations are substantial enough to warrant use of adaptive control, in view of the fact that a well designed classical feedback controller is insensitive to parameter values of the system to be controlled ([49, Horowitz]). This consideration is most likely due to the lack of confidence in using adaptive controllers since the raison d' être of adaptive control is precisely to deal with *modelled* uncertainties such as those in the parameters. How adaptive controllers deal with unmodelled uncertainties, i.e. the robustness issue, is a current research topic. The complementary, non-competing nature of adaptation and robustness should be noted; as Peter Caines remarked, one should learn as much as possible about the modelled part of the system and make it robust with respect to the behaviour of the unmodelled part.

The idea of an adaptive controller as a black box which can be placed in the feedback path of any system and result in a desired performance has been a dream in the minds of many [70, Wittenmark and Åström]. Compared with today's adaptive controller, a step closer to this dream might be the multi-adaptive controller depicted in Figure 6.1 below where higher levels of adaptation deal with further levels of ignorance about the system e.g. system structure. The nested adaptation structure, a generalization of the self-tuning controller of Figure 2.1, is likely to be extremely difficult to analyze for $n \geq 2$. The 1st adaptor is the most primitive one (dealing only with parameter uncertainty), and the information I_n needed to finally determine the control action depends on the flow of information $I_1 \rightarrow I_{n-1}$. The configuration in Fig. 6.1 embodies the all-important prerequisite of recursiveness and brings up an important point: What conceptual insight can on obtain concerning the fact that, for $n = 1$ (i.e. the case of theorem 2.5), the adaptor gives inaccurate information about the system parameters but, remarkably, results in asymptotically optimal performance? An answer is partly given by [53, Kumar and Praly] but, for a deeper understanding, it seems one must examine the relationship between the quality of information fed to the controller (or to an adaptor) and the performance of the overall adaptive scheme.

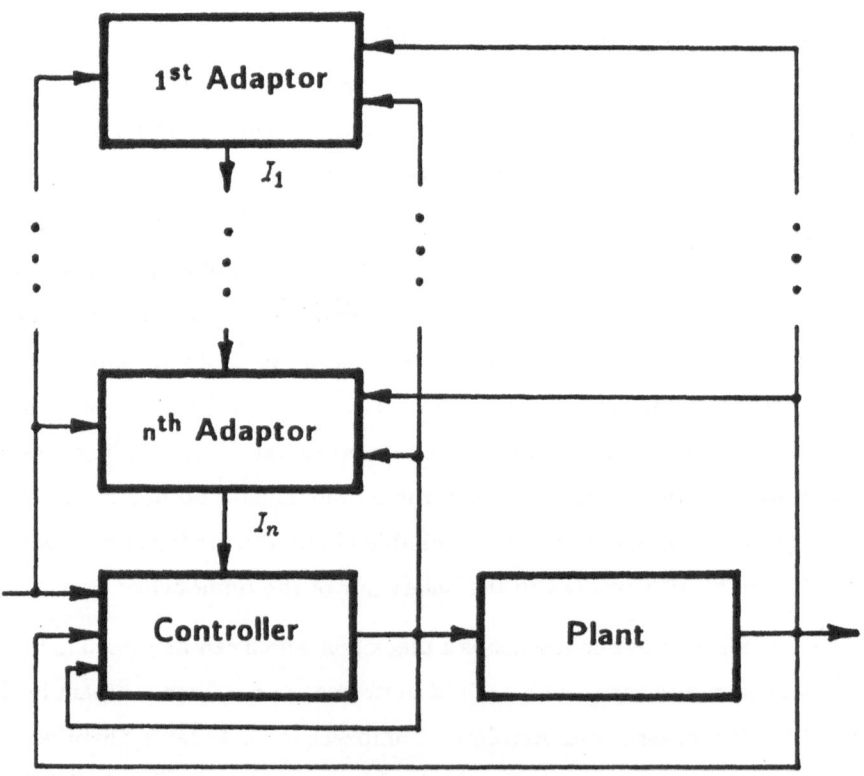

Figure 6.1   A Multi-Adaptive Controller

6.2   Summary of Results

In the body of this study, the theory of optimal (MV) control is carefully developed. Its motivation from the certainty equivalence principle is pointed out and the subtleties and differences found in various developments of the subject in the literature are explained. The two main types of parameter estimation schemes, namely, the stochastic approximation and the least squares schemes, as well as some of their variants, are outlined with some discussion of their origin. Minimum variance control and parameter estimation are then combined to yield a self-tuning controller which can stabilize an unstable system with sample mean squared bounded inputs, simultaneously causing it to track a prespecified deterministic reference sequence. Remarkably, consistent parameter estimation is not a prerequisite and in fact parameter convergence as such does not enter the analysis. However, consistent parameter estimation would be a desirable bonus and is shown to be achievable at the cost of sub-optimal tracking performance via the ideas of persistent excitation, sufficient richness, and continually disturbed controls. In certain cases, such as regulation about the zero reference point, convergence to a multiple of the true parameters can be established without these hypotheses.

The principal result of the experiments conducted is that the stabilizing and asymptotically optimal tracking properties of the adaptive controllers of theorems 2.5-2.7 are indeed displayed by simulation experiments designed so that the hypotheses of these algorithms are satisfied. The second result is that parameter convergence to their true values did not occur even when continually disturbed controls or a trajectory $\{y^*(t)\}$ satisfying certain technical hypotheses were used. Thirdly, concerning the output sample paths themselves, the Kolmogorov one-sample test performed on cross-sectional output data gave results indicating that the closed loop output process is asymptotically normally distributed.

Theoretical results on time varying systems are very limited. Certain results which explore the robustness of a self-tuning controller have been outlined and simulations have shown that the controller is in fact more robust with respect to parameter variations than theory predicts. Various considerations in designing adaptive controllers have been outlined and MV and ML controllers have been developed for an ARX system whose parameters are a finite state homogeneous Markov chain. Finally, a formulation

108

within which quite general stochastic systems (such as ARMAX systems with ARMA parameter processes) can be modelled using a Markov state process with stationary transition probabilities has been developed [21, Caines et.al.]. The L^2 stability of the state led to the existence of an invariant measure for the process and, via the ergodic theorem for Markov processes, to sample path convergence results of a function of the state.

It is believed that the aforementioned results illustrate the scope for effective new analytic approaches to adaptive control problems. A simple example was analyzed experimentally for which upper and lower bounds on the asymptotic average squared tracking error have been theoretically established. The experimental results substantiated the theoretical predictions and indicated that the theoretical results are perhaps rather conservative.

6.3 Future Prospects

Adaptive control is a difficult but highly motivated area of research, rich in ideas and concepts and full of unanswered questions. Research efforts have mainly concentrated on time invariant systems and the results have found a wide range of applications. The basic theme has been — and still is — the simultaneous on-line procedure of estimation (learning) and control. Asymptotically optimal performance has been established, but, surprisingly, added conditions are presently needed to establish consistent parameter estimation. Further, almost all results in adaptive control make use of some kind of positive real condition, something which has been criticized as artificial. However, at the expense of overparameterization, a solution is proposed in [64, Shah and Franklin]. Results on time varying systems have recently emerged and concentrated efforts in this area have revealed some promising links to artificial intelligence. The new approach to adaptive control via the L^2 stability of Markov processes and their associated invariant measures and simulations that were carried out have demonstrated the highly non-linear learning behaviour of such systems.

Concerning implementation, future research efforts should include comparative studies of the various types of controllers and provide possible classifications on when each is most useful accompanied with a critical comparison of performance. Concerning theory, the powerful techniques of Markov processes provides new hope for the development of results in the area of adaptive control of time varying systems. Indeed, this approach shows how the resolution of many problems in adaptive control may lie in the diversification of the mathematical techniques that are currently being used.

Appendix A.   Assumptions

Assumptions on the System Model

(M1)   The delay d is known, as well as upper bounds n', m' and ℓ' for n, m, and ℓ respectively.

(M2)   The delay d as well as n, m, and ℓ are known.

(M3) $B(z) = 0 \Rightarrow |z| > 1$.

(M4) $C^{-1}(z) - \frac{1}{2}$ is SPR.

(M5) $C(z) - \frac{1}{2}$ is SPR.

(M6) $C^{-1}(z)$ is SPR.

(M7) $z^{-1}(C(z) - A(z))$ and $B(z)$ have no common factors and their respective degrees $\max(n, \ell) - 1$ and m are known.

(M8) $A(z)$ and $B(z)$ have no common factors.

Assumptions on the Disturbance Sequence

(W1) $\dot{E}[w(t)|\mathcal{F}_{t-1}] = 0$ a.s. $\forall t \geq 1$.

(W2) $E[w^2(t)|\mathcal{F}_{t-1}] = \sigma_w^2 > 0$ a.s. $\forall t \geq 1$.

(W3) $E[w^2(t)|\mathcal{F}_{t-1}] = K_0(\omega)r^\varepsilon(t-1), \qquad 0 \leq \varepsilon(\omega) < 1$.

(W4) $\{w(t)\}$ and $\{v(t)\}$ are i.i.d. and mutually independent random processes with

$$E[w(t)] = E[v(t)] = 0 \text{ and } E[v^2(t)] = \sigma_v^2 > 0,$$

$$E[w^2(t)] = \sigma_w^2 > 0, \mathcal{F}_t = \sigma\{x_0, w(1) \cdots w(t), v(1) \cdots v(t)\}.$$

(W5) $\limsup_{N \to \infty} \frac{1}{N} \sum_{t=1}^N w^2(t) < \infty$ \qquad a.s.

(W6) $\lim_{N \to \infty} \frac{1}{N} \sum_{t=1}^N w^2(t) = \sigma_w^2(\omega) > 0$ \qquad a.s.

(W7) $\exists \delta > 0$ s.t. $\sup_{t \geq 1} E[\|w(t)\|^{2+\delta}|\mathcal{F}_{t-1}] < \infty$ \qquad a.s.

(W8)   The random quantities $x_0, w(1), w(2), \cdots$, are jointly absolutely continuous w.r.t. Lebesgue measure.

(W9) $E[w^2(t)|\mathcal{F}_{t-1}] \leq \sigma_w^2(\omega)$ a.s. $\forall\, t \geq 1$.

General Assumptions

(G1)   There exist random variables $T(\omega) > 0$, $\alpha(\omega) > 0$, and $\beta(\omega) > 0$ s.t.

$$\sum_{i=m(t)}^{m(t+\alpha)} \frac{\varphi(i)\varphi^T(i)}{r(i)} \geq \beta I \quad \forall t \geq T, \quad \forall \omega \,\epsilon\, \{\omega : \lim_{t \to \infty} r(t) = \infty\},$$

where $m(t) = \max\{n : p(n) \leq t\}$, $t \geq 0$, $p(n) = \sum_{i=0}^{n-1} \|\varphi(i)\|^2/r(i)$.

(G2) $E[u^2(t)] < K$ $\forall t$.

(T) $|y^*(t)| < \infty$ $\forall t \geq d$.

(MF1) If $\overline{H}(z)y_m(t) = 0$ where $\overline{H}(z) = 1 - \sum_{i=1}^{\overline{r}} \overline{h}_i z^i$, $\overline{r} > r$, then $\overline{h}_{r+1} = \cdots \overline{h}_{\overline{r}} = 0$.

(MF2) $H(z) = 0 \Rightarrow |z| = 1$ and $H(z_1) = H(z_2) = 0 \Rightarrow z_1 = z_2$.

Appendix B. Regression and Parameter Vectors

In what follows, $\bar{n} = \max(n, \ell - d + 1), \bar{n}' = \max(n', \ell' - d + 1)$.

(B1.1) $\varphi(\cdot), \hat{\vartheta}(\cdot) \in \Re^{\bar{n}'+m'+\ell'+d}, \qquad \vartheta^\circ \in \Re^{\bar{n}+m+\ell+d}$

$$\varphi(t-d) = [y(t-d) \cdots y(t-d-\bar{n}'+1), u(t-d) \cdots u(t-2d-m'+1),$$
$$- y^*(t-1) \cdots - y^*(t-\ell')]^T$$
$$\hat{\vartheta}(t) = [\hat{g}_0(t) \cdots \hat{g}_{\bar{n}'-1}(t), \widehat{(bf)}_0(t) \cdots \widehat{(bf)}_{m'+d-1}(t), \hat{c}_1(t) \cdots \hat{c}_\ell'(t)]^T$$
$$\vartheta^\circ = [g_0 \cdots g_{\bar{n}-1}, \widehat{(bf)}_0 \cdots \widehat{(bf)}_{m+d-1}, c_1 \cdots c_\ell]^T$$

(B1.2) $\varphi(\cdot), \hat{\vartheta}(\cdot) \in \Re^{\bar{n}'+m'+\ell'+d}, \qquad \vartheta^\circ \in \Re^{\bar{n}+m+\ell+d}$

$$\varphi(t-d) = [y(t-d) \cdots y(t-d-\bar{n}'+1), u(t-d) \cdots u(t-2d-m'+1),$$
$$- \bar{y}(t-1) \cdots - \bar{y}(t-\ell')]^T$$
$$\hat{\vartheta}(t) = [\hat{g}_0(t) \cdots \hat{g}_{\bar{n}'-1}(t), \widehat{(bf)}_0(t) \cdots \widehat{(bf)}_{m'+d-1}(t), \hat{c}_1(t) \cdots \hat{c}_{\ell'}(t)]^T$$
$$\vartheta^\circ = [g_0 \cdots g_{\bar{n}-1}, \widehat{(bf)}_0 \cdots \widehat{(bf)}_{m+d-1}, c_1 \cdots c_\ell]^T$$

(B2.1) $\varphi(\cdot), \hat{\vartheta}(\cdot) \in \Re^{n'+m'+\ell'+d}, \qquad \vartheta^\circ \in \Re^{n+m+\ell+1}$

$$\varphi(t-1) = [y(t-1) \cdots y(t-n'), u(t-1) \cdots u(t-m'-1),$$
$$(y(t-1) - y^*(t-1)) \cdots (y(t-\ell') - y^*(t-\ell'))]^T$$
$$\hat{\vartheta}(t) = [-\hat{a}_1(t) \cdots - \hat{a}_{n'}(t), \hat{b}_0(t) \cdots \hat{b}_{m'}(t), \hat{c}_1(t) \cdots \hat{c}_{\ell'}(t)]^T$$
$$\vartheta^\circ = [-a_1 \cdots - a_n, b_0 \cdots b_m, c_1 \cdots c_\ell]^T$$

(B2.2) $\varphi(\cdot), \hat{\vartheta}(\cdot) \in \Re^{n'+m'+\ell'+1}, \qquad \vartheta^\circ \in \Re^{n+m+\ell+1}$

$$\varphi(t-1) = [y(t-1) \cdots y(t-n'), u(t-1) \cdots u(t-m'-1),$$
$$(y(t-1) - \bar{y}(t-1)) \cdots (y(t-\ell') - \bar{y}(t-\ell'))]^T$$
$$\hat{\vartheta}(t) = [-\hat{a}_1(t) \cdots - \hat{a}_{n'}(t), \hat{b}_0(t) \cdots \hat{b}_{m'}(t), \hat{c}_1(t) \cdots \hat{c}_{\ell'}(t)]^T$$
$$\vartheta^\circ = [-a_1 \cdots - a_n, b_0 \cdots b_m, c_1 \cdots c_\ell]^T$$

(B3) $\varphi(\cdot), \hat{\vartheta}(\cdot) \in \Re^{\bar{n}+m+r+d}, \qquad 0 \le r \le \ell, \qquad \vartheta^\circ \in \Re^{\bar{n}+m+r+d}$

$$\varphi(t-d) = [y(t-d) \cdots y(t-d-\bar{n}+1), u(t-d) \cdots u(t-2d-m+1),$$
$$- y^*(t) \cdots - y^*(t-r+1)]^T$$

113

$$\widehat{\boldsymbol{\vartheta}}(t) = [\widehat{g}_0(t) \cdots \widehat{g}_{\overline{n}-1}(t), (\widehat{bf})_0(t) \cdots (\widehat{bf})_{m+d-1}(t), \widehat{\overline{g}}_0(t) \cdots \widehat{\overline{g}}_{r-1}(t)]^T$$

$$\boldsymbol{\vartheta}^\circ = [g_0 \cdots g_{\overline{n}-1}, (\widehat{bf})_0 \cdots (\widehat{bf})_{m+d-1}, \overline{g}_0 \cdots \overline{g}_{r-1}]^T$$

(B4) $\varphi(\cdot), \widehat{\boldsymbol{\vartheta}}(\cdot),$ and $\boldsymbol{\vartheta}^\circ \epsilon \, \Re^{\overline{n}+m+\ell+d+1}$

$$\varphi(t-d) = [y(t-d) \cdots y(t-d-\overline{n}+1), u(t-d) \cdots u(t-2d-m+1),$$
$$- y^*(t) \cdots - y^*(t-\ell)]^T$$

$$\widehat{\boldsymbol{\vartheta}}(t) = [\widehat{g}_0(t) \cdots \widehat{g}_{\overline{n}-1}(t), (\widehat{bf})_0(t) \cdots (\widehat{bf})_{m+d-1}(t), \widehat{\gamma}_0(t)\widehat{c}_1(t) \cdots \widehat{c}_\ell(t)]^T$$

$$\boldsymbol{\vartheta}^\circ = [g_0 \cdots g_{\overline{n}-1}, (\widehat{bf})_0 \cdots (\widehat{bf})_{m+d-1}, 1 \, c_1 \cdots c_\ell]^T$$

Appendix C. Miscellany

C1 Strict Positive Realness [17, Caines]

A polynomial $C : \mathbf{C} \to \mathbf{C}$ with real coefficients is strictly positive real (SPR) iff

$$C(e^{i\vartheta}) + C(e^{-i\vartheta}) > 0 \qquad \forall \vartheta \in [0, 2\pi].$$

It is true that if C is SPR so is C^{-1}.

C2 The Matrix Inversion Lemma [67, Sorenson]

Let m and n be arbitrary poisitive integers and suppose that $\mathbf{B} \in \Re^{n \times n}$ and $\mathbf{R} \in \Re^{m \times m}$ are symmetric and positive definite, while $\mathbf{H} \in \Re^{m \times n}$ is arbitrary. If

$$\mathbf{A} = \mathbf{B} - \mathbf{B}\mathbf{H}^T(\mathbf{H}\mathbf{B}\mathbf{H}^T + \mathbf{R})^{-1}\mathbf{H}\mathbf{B}$$

then $\mathbf{A}^{-1} = \mathbf{B}^{-1} + \mathbf{H}^T\mathbf{R}^{-1}\mathbf{H}$.

C3 Stochastic Transition Function [34, Doob]

Let X be a set of points x and $\mathcal{F}(X)$ a σ-algebra of sets in X. A function $p(:, \cdot)$ of $x \in X$ and $A \in \mathcal{F}(X)$ will be called a stochastic transition function if it has the following properties:

(i) $p(x, A)$ for fixed x determines a probability measure in A.

(ii) $p(x, A)$ for fixed A determines a function of x measurable w.r.t. $\mathcal{F}(X)$.

C4 Discrete Semidynamical System [63, Saperstone]

The pair (X, π) is called a discrete semidynamical system if X is a Hausdorff topological space and π is a mapping, $\pi : X \times Z^+ \to X$ which satisfies

(i) $\pi(x, 0) = x$ for each $x \in X$,

(ii) $-\pi(\pi(x, t), s) = \pi(x, t + s)$ for each $x \in X$ and $s, t \in Z^+$,

(iii) π is continuous.

C5 Invariant (Critical) Points of Semidynamical Systems [63, Saperstone]

The point x is called invariant (critical, equilibrium, or rest) point if $\pi(x, t) = x$ $\forall t \in Z^+$.

References

[1]   Anderson, B.D.O., and C. R. Johnson Jr. (1982). "Exponential Convergence of Adaptive Identification and Control Algorithms", *Automatica*, Vol. 18, No. 1, pp. 1-13.

[2]   Anderson, B.D.O., R.M. Johnstone (1983). "Adaptive Systems and Time Varying Plants", *Int. J. Control*, Vol. 37, No. 2, pp. 367-377.

[3]   Åström, K.J. (1983). "Theory and Applications of Adaptive Control — A Survey", *Automatica*, Vol. 19, No. 5, pp. 471-486.

[4]   Åström, K.J., P. Eykhoff (1971). "System Identification — A Survey", *Automatica*, Vol. 7, No. 2, pp. 123-162.

[5]   Åström, K.J., B. Wittenmark (1973). "On Self Tuning Regulators", *Automatica*, Vol. 19, No. 1, pp. 185-199.

[6]   Banman, A.L. (1985). "Simulation on an Industrial Design Tool", *Proceedings of the 11th Annual Advanced Control Conference*, Purdue University, pp. 85-92.

[7]   Bar-Shalom, Y., S.B. Gershwin (1978). "Applicability of Adaptive Control to Real Problems — Trends and Opinions", *Automatica*, Vol. 14, No. 4, pp. 407-408.

[8]   Bar-Shalom, Y., E. Tse (1974). "Dual Effect, Certainty Equivalence, and Separation in Stochastic Control", *IEEE Trans. Autom. Control*, Vol. AC-19, No. 5, pp. 494-500.

[9]   Becker, A.H., P.R. Kumar, C.Z. Wei (1985). "Adaptive Control with the Stochastic Approximation Algorithm: Geometry and Convergence", *IEEE Trans. Autom. Control*, Vol. AC-30, No. 4, pp. 330-338.

[10]   Beneš, V.E. (1967). "Existence of Finite Invariant Measures for Markov Processes", *Proc. Amer. Math. Soc.*, Vol. 18, No. 6, pp. 1058-1061.

[11]   Beneš, V.E. (1968). "Finite Regular Invariant Measures for Feller Processes", *J. Appl. Prob.*, Vol. 5, pp. 203-209.

[12]   Bierman, G.J. *"Factorization Methods for Discrete Sequential Estimation"*, Academic Press, 1977.

[13]   Boyd, S., S. Sastry (1983). "On Parameter Convergence in Adaptive Control",

Systems and Control Letters, Vol. 3, No. 6, pp. 311-319.

[14]   Burton, D. *"Introduction to Modern Abstract Algebra"*, Addison-Wesley, Reading, Massachusetts, 1967.

[15]   Caines, P.E. (to be published). *"Linear Stochastic Systems"*, John Wiley & Sons Inc.

[16]   Caines, P.E. (1981). "Stochastic Adaptive Control: Randomly Varying Parameters and Continually Disturbed Controls", *IFAC Congr. Proc.*, Kyoto, Japan, pp. 925-930.

[17]   Caines, P.E. (1980). "Passivity, Hyperstability and Positive Reality", *Proceedings of the Conference on Information Sciences and Systems*, Princeton, N.J., pp. 363-367.

[18]   Caines, P.E., H.F. Chen (1982). "On the Adaptive Control of Stochastic Systems with Random Parameters: A Counterexample", *Ricerche di Automatica*, Vol. 13, No. 1, pp. 190-196.

[19]   Caines, P.E., H.F. Chen (1985). "Optimal Adaptive LQG Control for Systems with Finite State Process Parameters", *IEEE Trans. Autom. Control*, Vol. AC-30, No. 2, pp. 185-189.

[20]   Caines, P.E., S. Lafortune (1984). "Adaptive Control with Recursive Identification for Stochastic Linear Systems", *IEEE Trans. Autom. Control*, Vol. AC-29, No. 4, pp. 312-321.

[21]   Caines, P.E., S. Meyn, A. Aloneftis (1986). "On the Stability of Markovian Systems with Applications to the Adaptive Control of Random Time Varying Systems", *2nd IFAC Workshop on Adaptive Systems in Control and Signal Processing*, Lund, Sweden, July 1986.

[22]   Cegrell, T., T. Hedqvist (1975). "Successful Adaptive Control of Paper Machines", *Automatica*, Vol. 11, No. 1, pp. 53-59.

[23]   Chen, H.F. (1981). "Quasi-Least-Squares Identification and its Strong Consistency", *Int. J. Control*, Vol. 34, No. 5, pp. 921-936.

[24]   Chen, H.F. (1982). "Self-Tuning Controller and its convergence under correlated noise", *Int. J. Control*, Vol. 35, No. 6, pp. 1051-1059.

[25]   Chen, H.F. (1982). "Strong Consistency in System Identification under corre-

lated noise", *Proc. 6th IFAC Symposium on Identification and System Parameter Estimation*, Vol. 2, pp. 964-969.

[26]   Chen, H.F. (1982). "Strong Consistence and Convergence Rate of Least Squares Identification", *Scientia Sinica (Series A)*, Vol. 25, No. 7, pp. 771-784.

[27]   Chen, H.F. (1984). "Recursive System Identification and Adaptive Control by Use of the Modified Least Squares Algorithm", *SIAM J. Control and Optimization*, Vol. 22, No. 5, pp. 758-776.

[28]   Chen, H.F., P.E. Caines (1983). "Adaptive Control and Identification for Stochastic Systems with Random Parameters", *Proceedings of the IFAC Workshop on Adaptive Systems in Control and Signal Processing*, San Fransisco, California, USA, pp. 179-184.

[29]   Chen, H.F., P.E. Caines (1985). "The Strong Consistency of the Stochastic Gradient Algorithm of Adaptive Control", *IEEE Trans. Autom. Control*, Vol. AC-30, No. 2, pp. 189-192.

[30]   Chen, H.F., P.E. Caines (1985). "On the Adaptive Control of a Class of Systems with Random Parameters and Disturbances", *Automatica*, Vol. 21, No. 6, pp. 737-741.

[31]   Chen, H.F., L. Guo (1985). "Strong Consistency of Parameter Estimates for Discrete-Time Stochastic Systems", *J. Sys. Sci. & Math. Scis.*, Vol. 5 (2), pp. 81-93.

[32]   Chen, H.F., L. Guo (1985). "Adaptive Control with Recursive Identification for Stochastic Linear Systems", *Technical Report*, To appear in Control and Dynamic Systems, C.T. Leondes (ed.), Academic Press.

[33]   Cordero, A.O., D.Q. Mayne (1981). "Deterministic Convergence of a Self Tuning Regulator with Variable Forgetting Factor", *IEEE Proc.*, Vol. 128, Part D, No. 1, pp. 19-23.

[34]   Doob, J.L. *"Stochastic Processes"*, John Wiley & Sons, Inc., 1953.

[35]   Dumont, G.A., P.R. Bélanger. "Self Tuning Control of a Titanium Dioxide Kiln", *IEEE Trans. Autom. Control*, Vol. AC-23, No. 4, pp. 532-538.

[36]   Dvoretzky, A. (1956). "On Stochastic Approximation", *Proc. 3rd Berkeley Sym. on Math. Stat. and Prob.*, J. Neyman (ed.), Berkeley, University of California

Press, pp. 39-55.

[37]   Feldbaum, A.A. (1960-61). "Dual Control Theory I - IV", *Automation and Remote Control*, Vol. 21, No. 9, pp. 874-880, No. 11, pp. 1033-1039, Vol. 22, No. 1, pp. 1-12, No. 2, pp. 109-121.

[38]   Foguel, S.R. (1962). "Existence of Invariant Measures for Markov Processes", *Proc. Amer. Math. Soc.*, Vol. 13, No. 6, pp. 833-838.

[39]   Fortescue, T.R., L.S. Kershenbaum, B.F. Ydstie (1981). "Implementation of Self-Tuning Regulators with Variable Forgetting Factors", *Automatica*, Vol. 17, No. 6, pp. 831-835.

[40]   Gilbart, J.W., G.C. Winston (1974). "Adaptive Compensation for an Optical Tracking Telescope", *Automatica*, Vol. 10, No. 2, pp. 125-131.

[41]   Goodwin, G.C., P.J. Ramadge, P.E. Caines (1981). "Discrete Time Stochastic Adaptive Control", *SIAM J. Control and Optimization*, Vol. 19, No. 6, pp. 829-853. "Corrigendum", Vol. 20, No. 6, November 1982, pg. 893.

[42]   Goodwin, G.C., P.J. Ramadge, P.E. Caines (1980). "Discrete Time Multivariable Adaptive Control", *IEEE Trans. Autom. Control*, Vol. AC-25, No. 3, pp. 449-456.

[43]   Goodwin, G.C., K.S. Sin. *"Adaptive Filtering Prediction and Control"*, Prentice-Hall Inc., 1984.

[44]   Goodwin, G.C., K.S. Sin, K.K. Saluja (1980). "Stochastic Adaptive Control and Prediction — The General Delay-Coloured Noise Case", *IEEE Trans. Autom. Control*, Vol. AC-25, No. 5, pp. 946-950.

[45]   Goodwin, G.C., E.K. Teoh (1985). "Adaptive Control of a Class of Linear Time Varying Systems", *Proceedings of the IFAC Workshop on Adaptive Systems in Control and Signal Processing*, San Francisco, California, USA, pp. 1-6.

[46]   Hägglund, T. (1983). "The Problem of Forgetting Old Data in Recursive Estimation", *IFAC Workshop on Adaptive Systems in Control and Signal Processing*, June 20-22, San Fransisco, California, pp. 213-214.

[47]   Harris, C.J., S.A. Billings (editors). *"Self-Tuning and Adaptive Control: Theory and Applications"*, Peter Peregrinus Ltd., U.K., second edition, 1985.

[48]   Hopf, E. (1959). "The General Temporal Discrete Markoff Process", *J. Rational*

Mechanics and Analysis, Vol.3, pp. 13-45.

[49] **Horowitz, I.** *"Synthesis of Feedback Systems"*, Academic Press, 1963.

[50] **Isermann, R., K. H. Lachmann** (1985). "Parameter-Adaptive Control with Configuration Aids and Supervision Function", *Automatica*, Vol. 21, No. 6, pp. 625-638.

[51] **Jacoby, S.L.S., J.S. Kowalik, J.T. Pizzo.** *"Iterative Methods for Nonlinear Optimization Problems"*, Prentice-Hall Inc., 1972.

[52] **Kershenbaum, L.S., T.R. Fortescue** (1981). "Implementation of on-line Control in Chemical Process Plants", *Automatica*, Vol. 17, No. 6, pp. 777-788.

[53] **Kumar, P.R., L. Praly** (1985). "Self-Tuning and Convergence of Parameter Estimates in Minimum Variance Tracking and Linear Model Following", *Tech. Report* 85-518, University of Illinois at Champaign-Urbana, Indiana, U.S.A.

[54] **Lafortune, S.** (1982). "Adaptive Control with Recursive Identification for Stochastic Linear Systems", *M. Eng. Thesis*, McGill University, Montreal, Canada.

[55] **Ljung, L., B.D.O. Anderson** (1984). "Adaptive Control, Where Are We?", *Automatica*, Vol. 20, No. 5, pp. 499-500.

[56] **Ljung, L., T. Söderström.** *"Theory and Practice of Recursive Identification"*, The MIT Press, Cambridge Massachusetts, 1983.

[57] **Meyn. S.P., P.E. Caines** (1984). "The Zero Divisor Problem of Multivariable Stochastic Adaptive Control", Systems and Control Letters, Vol. 6, No. 4, pp. 235-238.

[58] **Meyn, S., P.E. Caines** (1986). "A New Approach to Stochastic Adaptive Control", To appear in *Proceedings of the 25th IEEE Conference on Decision and Control*, December 1986, Athens, Greece.

[59] **Narendra, K.S., R.V. Monopoli (eds.).** *"Applications of Adaptive Control"*, Academic Press, 1980.

[60] **NASA Flight Research Center** (1971). "Experience with the X-15 Adaptive Flight Control System", *NASA* TN-D-6208.

[61] **Patchell, J.W., O.L.R. Jacobs** (1971). "Separability, Neutrality, and Certainty Equivalence", *Int. J. Control*, Vol. 13, No. 2, pp. 337-342.

[62] Robbins H., S. Monro (1951). "A Stochastic Approximation Method", *Ann. Math. Stat.*, Vol. 22, pp. 400-407.

[63] Saperstone, S.H. *"Semidynamical Systems in Infinite Dimensional Spaces"*, Springer Verlag New York Inc., 1981.

[64] Shah, H., G.F. Franklin (1982). "On Satisfying Strict Positive Real Condition for Convergence by overparameterization" *IEEE Trans. Autom. Control*, Vol. AC-27, No. 3, pp. 715-716.

[65] Sin, K.S., G.C. Goodwin (1982). "Stochastic Adaptive Control Using a Modified Least Squares Algorithm", *Automatica*, Vol. 18, No. 3, pp. 315-321.

[66] Solo, V. (1979). "The Convergence of AML", *IEEE Trans. Autom. Control*, Vol. AC-24, No. 6, pp. 958-963.

[67] Sorenson, H.W. *"Parameter Estimation"*, Marcel Dekker Inc., 1980.

[68] Tsypkin, Ya.Z. *"Adaptation and Learning in Automatic Systems"*, Academic Press, 1971.

[69] Wittenmark, B. (1975). "Stochastic Adaptive Control Methods: A Survey", *Int. J. Control*, Vol. 21, No. 5, pp. 705-730.

[70] Wittenmark, B., K.J. Åström (1984). "Practical Issues in the Implementation of Self-Tuning Control", *Automatica*, Vol. 20, No. 5, pp. 595-605.

Lecture Notes in Control and Information Sciences

Edited by M. Thoma and A. Wyner

Lecture Notes in Control and Information Sciences

Edited by M. Thoma and A. Wyner

Lecture Notes in Control and Information Sciences

Edited by M. Thoma and A. Wyner

Vol. 98: A. Aloneftis
Stochastic Adaptive Control
Results and Simulations
XII, 120 pages, 1987.